T0137050

Studies in Systems, Decision and Control

Volume 165

Series editor

Janusz Kacprzyk, Polish Academy of Sciences, Warsaw, Poland
e-mail: kacprzyk@ibspan.waw.pl

The series "Studies in Systems, Decision and Control" (SSDC) covers both new developments and advances, as well as the state of the art, in the various areas of broadly perceived systems, decision making and control- quickly, up to date and with a high quality. The intent is to cover the theory, applications, and perspectives on the state of the art and future developments relevant to systems, decision making, control, complex processes and related areas, as embedded in the fields of engineering, computer science, physics, economics, social and life sciences, as well as the paradigms and methodologies behind them. The series contains monographs, textbooks, lecture notes and edited volumes in systems, decision making and control spanning the areas of Cyber-Physical Systems, Autonomous Systems, Sensor Networks, Control Systems, Energy Systems, Automotive Systems, Biological Systems, Vehicular Networking and Connected Vehicles, Aerospace Systems, Automation, Manufacturing, Smart Grids, Nonlinear Systems, Power Systems, Robotics, Social Systems, Economic Systems and other. Of particular value to both the contributors and the readership are the short publication timeframe and the world-wide distribution and exposure which enable both a wide and rapid dissemination of research output.

More information about this series at http://www.springer.com/series/13304

Mohamed Elhoseny · Aboul Ella Hassanien

Dynamic Wireless Sensor Networks

New Directions for Smart Technologies

Mohamed Elhoseny
Faculty of Computers and Information
Mansoura University
Dakahlia
Egypt

Aboul Ella Hassanien
Department of Information Technology
Cairo University
Giza
Egypt

ISSN 2198-4182 ISSN 2198-4190 (electronic)
Studies in Systems, Decision and Control
ISBN 978-3-030-06521-8 ISBN 978-3-319-92807-4 (eBook)
https://doi.org/10.1007/978-3-319-92807-4

Printed on acid-free paper

This Springer imprint is published by the registered company Springer International Publishing AG part of Springer Nature
The registered company address is: Gewerbestrasse 11, 6330 Cham, Switzerland

Foreword

It was 2014 when Dr. Mohamed Elhoseny joined the Computer Vision and Intelligent Systems lab at the University of North Texas, Denton, Texas, USA, where I have served as the director since 2006. Our research interests align in the field of artificial intelligence, in particular, optimization methods for wireless sensor network. Dr. Elhoseny is a rising star researcher in the field of wireless sensor network and, to my best knowledge, Dr. Aboul Ella Hassanien is an established researcher in this field. This book is a pioneering effort on the dynamic wireless sensor networks, which focuses on the smart applications. In-depth discussions on the security measurements, large volume of data, and intelligent analysis shed new lights on the future research and development of sensor networks in the era of Internet-of-things and big data. The readers will find extraordinary values in its visionary words.

Denton, Texas
January 2018

Xiaohui Yuan

Preface

Recently, Wireless Sensor Networks (WSNs) have gained great attention due to their ability to work effectively in different working fields. It has been widely used in health care, transportation management, military surveillance, etc. Battery-operated sensors are placed in open fields without human attendance and acquire data continuously over time. However, collecting and processing data in smart and modern WSN-based applications such as intelligent transportation systems is still a big challenge because of the limitations of its resources. There are great efforts to apply intelligent models in WSNs to collect, transmit, and process data especially in real-time applications such as SCADA systems. While the society of WSN focuses more on the designs of sensing, event-handling, data-retrieving, communication, and coverage issues, the society of the other Pervasive and Mobile Computing, i.e., SCADA systems, focuses more on the development of cross-domain intelligence from multiple WSNs and the interactions between the virtual world and the physical world. In such live data monitoring environments, systems face significant threats and attacks. After an attack occurred, a system requires forensic investigation to understand the cause and effects of the intrusion or disruption of the systems services. Despite the great efforts to acquire and process large amount of live data on dynamic WSN-based systems, the continuous change of this type of data and the risk on the systems services make it a big challenge. With adequate preparation and the appropriate response planning and execution, it is possible to successfully perform a secure data processing using WSN in such harsh environments. Moreover, the increasing usage of WSNs as a communication medium for recent commercial applications made them more overloaded with big amount of data. Thus, we believe that the near future scientific research will focus on adapting dynamic WSNs for intelligent and smart applications.

This book aims to provide a collection of high-quality research works that address broad challenges in both theoretical and application aspects of dynamic WSNs for intelligent and smart applications in different environments. This book contains a set of book chapters that stimulates the continuing effort on the application of the intelligent WSNs model that leads to solve the problem of data processing in a limited resource WSN-based environment.

This book would not have been possible without the support of many people. We wish to express our gratitude to Prof. Xiaohui Yuan who was abundantly helpful and offered invaluable assistance, support, and guidance. Deepest Gratitude is also due Scientific Research Group in Egypt (SRGE) without its members' cooperation and assistance this book would not has been successful. Finally, Dr. Elhoseny wishes to express his thanks to his family and his wife Dr. Noura Metawa for their efforts and continuous support.

Egypt
January 2018

Mohamed Elhoseny
Aboul Ella Hassanien

Contents

Acronyms

[DDEEC]	Developed Distributed Energy-Efficient Clustering
ABC	Artificial Bee Colony
AC	Address-centric routing
ACO	Ant Colony Optimization
AES	Advanced Encryption System
AR-SC	Adjustable Range Set Covers
BS	Base Station
CH	Cluster Head
DC	Data-Centric Model
DE	Differential Evolution
DOS	Denial-of-Service
ECC	Elliptic Curve Cryptography
FDA	Flow Decomposition Algorithm
FDSSP	Fixed Directional Sensor Scheduling Problem
FDSSP-NUE	Fixed Directional Sensor Scheduling Problem-Non-uniform initial Energy version
FDSSP-UE	Fixed Directional Sensor Scheduling Problem-Uniform initial Energy version
FL	Fuzzy Logic
FSA	Fish Swarm Algorithm
GA	Genetic Algorithm
GABEEC	Genetic Algorithm-Based Energy-Efficient Clusters
GASONeC	Genetic Algorithm-based Self-organizing Network Clustering
GA-WCA	Genetic Algorithm and Weighed Clustering Algorithm
HE	Homomorphic Encryption
HEED	Hybrid Energy-Efficient Distributed
IP	Integer Programming
IDS	Intrusion Detection System
KDS	Key Distribution Server
LEACH	Low Energy Adaptive Clustering Hierarchy

LELE	Leader Election with Load balancing Energy
LNCA	Local Negotiated Clustering Algorithm
LND	Local Node Density
MSC	Maximum Set Covers
NN	Neural Networks
PSO	Particle Swarm Optimization
ROI	Region of Interest
ROI	Reinforcement Learning
ROM	Read Only Memory
SEP	Stable Election Protocol
SGA	Stud Genetic Algorithm
SLEACH	Secure Low Energy Adaptive Clustering Hierarchy
SN	Sensor Node
SPL	Security Procedure Layer
TSEP	Threshold Sensitive Stable Election Protocol
VP-NL	Variable Power Network Lifetime
WSN	Wireless Sensor Network

Symbols

S-CH	Secure CH Selection
S-CF	Secure Cluster Formation
S-DA	Secure Data Aggregation
S-DR	Secure Data Routing
A_i	Attack Identifier, i.e., A_1 means DOS attack
M	The required memory size
E	The energy consumption ratio
P	The required processing time / complexity of computations
D	Dynamic clustering
S	Static clustering
$E_s^A(l)$	The energy used to acquire l bits of data
$E_s^P(l')$	The energy used to process l' bits of data
s	Sensor node
$E_s^R(l'')$	The energy used to receive l'' bits of data
$E_s^T(l',d)$	The energy of transmitting l' bits of data over a distance d
E_s^T	The transmitter energy consumption
E_s^R	Receiver energy consumption
E_i	The idle energy expenditure
t	The time / the round ID
$E(0)$	The initial energy of the node and t
N_s	The number of SN inside a cluster
E	The constant energy expenditure that includes energy for data acquisition, processing, and idle
$D(s',s)$ and $D(s,B)$	Give the distances between nodes s' and the CH s and from node s to the BS (B), respectively
δ	A neighborhood distance threshold
$\|\cdot\|$	A function that gives the set size
$\mathbb{D}$	The network communication distance
C	The number of clusters in a network

$\tilde{E}$	The total energy cost if the messages are transmitted directly from all nodes to the BS
$q \in [1, Q]$	The number of generations
P	The population size
U	The pool of chromosome
$\tilde{U}$	An intermediate pool of chromosomes
α	The crossover probability
β	The mutation probability
$g(n)$	The cost to reach the CH
$h(n)$	The cost to get from the CH to the BS
$\|H_i, n\|$	The count of hindrances between two nodes i and n
ED	The energy aware distance
$\rho(x, BS)$	The path from the CH x to the BS
$E_s^M(x, y)$	The consumed energy to move from location x to location y
p	A base point that lies on an elliptic curve
k_A	The private key of node A
$\bar{k}_A$	The public key of ode A
R	A shared secret key
p_1 and p_2	The encryption key parts
$E()$	The encryption function
c_i	A cipher text
M	A group of messages before encryption
C	A group of messages after encryption

List of Figures

List of Tables

Part I
WSN for Complex and Mobile-Based Applications

The first part of this book discusses a set of complex and mobile-based applications of wireless sensor networks. This part aims to show how a WSN affects the efficiency of each of those applications and how its performance is affected by the working environment. It introduces briefly the wireless sensor network concepts and terminologies such as the network lifetime, terminologies, role in real life, and its structuring models. Then, it explains the different routing models and network structures for data routing in WSN. Moreover, the application areas of WSN and the network types, i.e., homogeneous and heterogeneous network, are explained. One of the main goals of this part is to provide extended experiments in different environments with a variety of parameters to evaluate WSNs performance.

Chapter 1
Mobile Object Tracking in Wide Environments Using WSNs

Abstract Covering a specific field and transferring data to Base Station (BS) is a real defiance. Although there are extended efforts to build a routing protocol that avoids a high energy consumption, the dynamic nature and complex environments of most of WSN recent application makes building such protocol a big challenge. To avoid energy exhaustion, many machine learning algorithms are used to manage the network operations. We proposed a new model to optimize the coverage requirements in WSNs to provide continuous monitoring of specified targets for longest possible time with limited energy resources. Moreover, we allow sensor nodes to move to appropriate positions to collect environmental information. The proposed model is based on the continuous and variable speed movement of mobile sensors to keep all targets under their cover all times. To further prove that the proposed model is better than other related work, a set of experiments in different working environments and a comparison with the most related work are conducted.

1.1 Introduction

Wireless Sensor Networks (WSNs) are widely used in many applications such as industry [1–10], environmental monitoring [11–14], health care [15–22], and agriculture [23–29]. Due to the limited energy, a sensor can remain active only for a finite amount of time. Thus, sensors are organized into different groups, namely sensor cover, in such a way that each cover monitors the targets for a certain duration, and the optimal use of the sensors increases the sensor network lifetime [30]. This motivates the deployment of redundant sensors to cover the area of interest and to organize the sensors to prolong the coverage time after deployment. This problem is the K-coverage problem, which requires a minimum of k sensor nodes to monitor one target [31–34]. Intelligent Algorithms are extensively used for WSN to avoid energy consumption. Genetic Algorithm (GA) [35] is a widely used optimization algorithm, and it has been used frequently in different applications especially in WSN. A sensor network is some tiny sensor nodes of low costs that cover a certain Region of Interest (ROI) to measure data using different sensing capabilities and transmit it to the base station (BS). To minimize power consumption in data transmission, it is preferable

M. Elhoseny and A. E. Hassanien, *Dynamic Wireless Sensor Networks*, Studies in Systems, Decision and Control 165, https://doi.org/10.1007/978-3-319-92807-4_1

to use multi-hop transmission [36] to reach the BS instead of direct transmission, especially in large ROIs with only one BS.

Many target coverage methods assume that the targets are known, and each target is covered by one sensor [37–39]. However, these algorithms have a serious drawback when a sensor runs out of energy. Hence, covering each target with more than one sensor at a time provides a more robust solution [40, 41]. In [41], the flow decomposition algorithm (FDA) was introduced and compared with Fixed Directional Sensor Scheduling Problem (FDSSP) that was proposed in [42]. FDA aims to decompose the maximum flow into a set of single flows, and every single flow represents a source to a sink path. The sensors of that path form a cover, and the amount of flow passing through this path is equal to the lifetime of this cover. The FDSSP seeks a fixed directional sensor schedule which maximizes the lifetime. Factors such as sensor network topology, sensor activation mode, and sensor role must be taken into consideration for identifying an optimal network management solution [38, 39, 41, 43]. In FDSSP, given a set of fixed directional sensors which have already been placed. In [42], two versions of FDSSP were introduced. The first one was the Uniform initial Energy version (FDSSP-UE) in which all the sensor nodes are assumed to have the same initial energy. This version has many problems as reported in [42]. The second version was the Non-Uniform initial Energy version (FDSSP-NUE) in which different nodes may have different initial energy; and this version, i.e., FDSSP-NUE, was used for solving the problems of the FDSSP-UE. The Variable Power Network Lifetime (VP-NL) scheduling scenario was proposed in [44]. In this scenario, it was assumed that each sensor could modulate its sensing range by dynamically varying its operating power, e.g., radar sensors. In VP-NL, A polynomial algorithm was proposed, and many experiments and numerical simulations were conducted to show its effectiveness. With the proliferation of sensors, a wireless sensor network is no longer stationary, which greatly expands the applications such as tracing animal movements applications [26, 45, 46] and environmental monitoring [47], in contrast to the stationary sensor networks. In many cases, monitoring the whole area might be unnecessary, especially if the dynamic nature of the observed processes is taken into account [48]. When sensors are equipped with motion capabilities, monitoring some points of interest instead of the whole area increases the network performance and permits time-dependent coverage. In a mobile sensor network, combining target coverage with the connectivity of sensors to the data sink is still an open challenge [49, 50].

Non-stationary K-coverage is often needed when a reliable monitoring capability is desired as in surveillance and military applications. Due to the energy constraint of wireless sensors and often infeasibility of replacement or recharging, it is necessary for the sensors to be densely deployed. Keeping all sensors active will deplete their energy quickly. A typical scenario is multi-agent based corporative field monitoring. Mobile agents collect and transform data to ensure integrity and security [51–55] in the parameter.

We propose a Genetic Algorithm (GA) [56, 57] based method to optimize the coverage in WSNs to monitor specified targets for the longest possible time with limited energy. The sensor nodes are non-stationary and can move in the field to collect data. We do not assume the mobility speed of the sensors that can be at continuous or variable speed. GA has been applied to WSN [58–62]. In the introduced problem, the data transmission round is a period that the data of targets are collected and transmitted to the base station. A GA-based method was proposed to optimize the sensor covers with a goal of maximizing the network lifetime by determining the mode of sensor covers. Based on a set of factors such as the coverage range of each sensor, expected consumed energy, the distance to the base station, and targets positions, the GA forms the covers after determining the optimum cover heads that are responsible for transferring the data to the base station. Thus, the proposed model ensures that the monitored area is fully covered by a minimum number of sensors.

This study has two main contributions. Firstly, GA-based cover forming method that creates all possible sensor covers. Secondly, a WSN covers management method that switches between different sensor covers to maximize the network lifetime. Section 1.2 summarizes the related work of the target covering problem. Section 1.3 explains the proposed model in details. It discusses the mathematical model and the proposed algorithm for the target coverage problem. Section 1.4 summarizes the experimental results and discussions of the experiments. Finally, chapter summary is presented in Sect. 1.5.

1.2 Related Work

Target covering problem has been attracting significant attention in WSNs [63–65]. In [65], a heuristic algorithm that selects mutually exclusive sets of sensor nodes was proposed. The members of a set cover the whole area completely, and only one of the sets is active at any time. This algorithm achieved a significant energy saving while fully preserving coverage. Cardi et al. proposed a method to extend the lifetime of the sensor network by reorganizing the sensors into a maximal number of disjoint set covers [64]. Moreover, the sensors from the current active set are utilized for (1) monitoring all targets and (2) transmitting the collected information, while the nodes from all the other sets are in sleep mode. The method [64] achieved competitive results, and it outperformed the algorithm in [65] regarding the increased number of produced disjoint sensor covers. Berman et al. [63] introduced a power efficient monitoring model which proposed: (1) an efficient data structure to efficiently represent the monitoring area; (2) an algorithm for sensor monitoring; and (3) distribution protocols to make a balance between the monitoring and communication power consumption. The results of their proposed model revealed a significant advantage in quality, flexibility, and scalability.

Cardei et al. in [66] improved the model that was introduced in [64] by increasing the lifetime of the network without the constraint that chosen set covers are disjoint; thus, a sensor may appear in different covers. The target covering was modeled as a Maximum Set Covers (MSC) problem, and two heuristic algorithms were employed to compute the sets based on linear programming and greedy approach.

Most studies in the field of wireless sensor network have assumed that the sensors have the same sensing range [64–66]. On the other hand, in [67], the sensors assumed to have adjustable sensing ranges; hence, the target covering problem was formulated as an Adjustable Range Set Covers (AR-SC) problem. Three heuristic algorithms were introduced to solve the AR-SC problem, and the goal was to maximize the numbers of set covers for the ranges associated with each sensor. One of the three algorithms was based on the Integer Programming (IP), and the other two algorithms were based on greedy approach. In different studies, each active sensor covered all targets in its sensing region [64–68], whereas Liu et al. reported that each sensor covered only one target at a time, and the sensor can freely select its target to cover it. The optimal solution was finding the target observation schedule that maximizes the network lifetime [69]. The findings of this work were enhanced in [70], where each target was covered by at least K sensors, this is called K-coverage.

There are many studies that employ the bio-inspired algorithms to optimize the K-coverage problem. In [71], the Particle Swarm Optimization (PSO) and Simulated Annealing (SA) algorithms were combined for energy-efficient coverage in WSNs. In another research, the Artificial Bee Colony (ABC) algorithm was used for finding the optimal deployment positions in a three-dimensional terrain [72]. Fish Swarm Algorithm (FSA) was utilized for coverage optimization of WSNs [73]. Wang et al. in [74], optimized the K-coverage problem using the Biogeography Based Optimization (BBO) algorithm, and they compared the performance of BBO with the ABC algorithm and Stud Genetic Algorithm (SGA). Their results demonstrated that the BBO algorithm yielded results better than both ABC and SGA algorithms. The PSO was combined with the Differential Evolution (DE) algorithm (PSO+DE) in [75] to optimize the K-coverage problem, and the results proved that the PSO+DE algorithm increased the lifetime of the network by optimizing the coverage of the sensors in comparison with the standard PSO algorithm. GA has been used to extend the life of WSNs [58, 60]. Ebrahimian et al. employed a GA-based model for K-coverage in WSN [58]. The GA aims to find K-coverage states that minimize the number of on-duty sensors. Mnasri et al. utilized the GA to search for sensor nodes in a WSN to (1) maximize the coverage area, and (2) optimize the audio localization in wireless sensor networks [43].

1.3 The Problem Formulation and the Proposed Solution

Let $N = (T, S)$ be a wireless sensor network, where $S = \{s_1, s_2, \ldots, s_m\}$ is a set of sensors with a sensing range R_s, $T = \{t_1, t_2, \ldots, t_n\}$ is a set of targets with known locations, m is the number of sensors, and n is the number of targets. Each target is sensed with one or more sensors, e.g. the target t_1 is covered with the sensors s_1 and s_3. The collected data are processed by a sink node. A sensor is in the *active* mode if it acquires or relays data, or both. A sensor in the *sleep* state when the sensor is not performing any tasks.

The network lifetime [76, 77] is defined as the period from the network being set up till (1) one or more targets cannot be covered by at least one sensor, or (2) a route between each sensor to the sink cannot be found. The network lifetime is maximized in the Connected Target Coverage (CTC) problem, which can be modeled as a Maximum Cover Tree (MCT).

The proposed model identifies the maximum number of non-disjoint sets of the sensors, namely sensor cover, which is bounded by C_{max} such that at a given point of time all targets are monitored, and only one sensor in a cover is active. Let C be the set of sensor covers, i.e., $C = \{C_1, \ldots, C_{max}\}$. Each sensor cover, C_k, $C_k \in C$, is enough to cover all the targets in the network. The lifetime of a sensor cover C_k is denoted by $X(C_k)$, and it cannot exceed the remaining energy of a sensor in C_k, which has the minimum lifetime, i.e., $X(C_k) = min\ [C_k(b_i)]$. The objective of the target coverage problem is to generate a maximum number of sensor covers to prolong the network lifetime.

The energy-efficient target coverage problem is formulated as a maximization problem that aims to maximize the total lifetime of all sensor covers, i.e., $\sum_k X(C_k)$), (we shall refer to x_k) as indicated in Eq. (1.1). Hence, the goal of target coverage problem is to find the complete family of sensor covers which has the maximum aggregated network lifetime among all the families of sensor covers.

$$\begin{aligned} \text{Maximize} \quad & \sum p x_p \\ \text{subject to} \quad & \\ & \sum_p B_{ip} x_p \leq b_i (\forall s_i) \\ & x_p \geq 0 (\forall C_p) \end{aligned} \tag{1.1}$$

where p represents the index of sensor covers, x_p is the sensor cover, s_i is the sensor node with index i, b_i indicates the lifetime for the sensor s_i, C_p is the sensor cover with index p, and B is constant for the constraint matrix and it is defined as:

$$B_{ij} = \begin{cases} 1 \text{ if } s_i \in C_p \\ 0 \text{ otherwise} \end{cases} \tag{1.2}$$

The GA goal of the proposed model is to maximize the network lifetime [78]. The framework of the proposed model is shown in Fig. 1.1. As shown, the proposed model consists of three main phases. In the first phase, is called *encoding* phase, the system is initialized, i.e., the sensor nodes are distributed in a working field, and then a binary chromosome is used to encode the sensor nodes within the field. In the second phase, is called *optimization* [79–85] phase, the GA randomly generates a set of chromosomes that forms its initial population. Dependently, GA algorithm runs after each round to choose the optimum number of cover heads (represented by 1). Depending on the sensing range of each sensor and the targets positions, covers will be formed. In the third phase, is called *validation* phase, each chromosome is

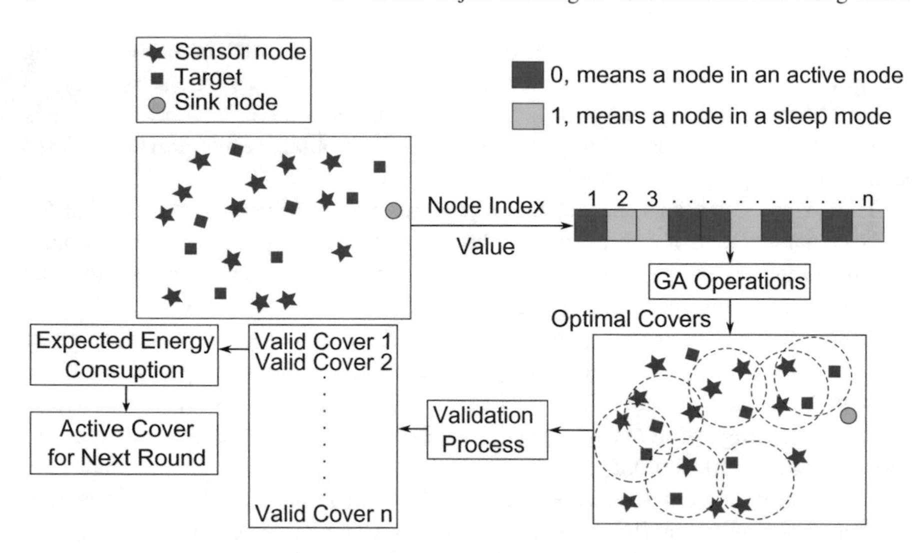

Fig. 1.1 The framework of the proposed model

evaluated to make sure that all targets are covered. By getting all possible covers, the expected consumed energy of each proposed cover is calculated to determine which one will be active for the coming round.

1.3.1 Cover Head Selection Factors

In the search for the suitable cover heads, the following network properties are considered:

- The distance between a cover head and the base station,
- Remaining battery power, and
- Expected consumed energy.

Simultaneously, the expected consumed energy is considered to choose the active cover for the next round. The consumed energy E in a cover consists of the total consumed energy for all of its sensors. The consumed energy E_s of a normal sensor node s is the summation of energy used to:

1. acquire l bits of data $(E_s^A(l))$,
2. receive l' bits of data $(E_s^R(l'))$,
3. process l'' bits of data $(E_s^P(l''))$,
4. transmit l'' bits of data over a distance d $(E_s^T(l'', d))$, and
5. move from location x to location y.

$$E_s = E_s^A(l) + E_s^R(l') + E_s^P(l'') + E_s^T(l'', d) + E_s^M(x, y), \tag{1.3}$$

where $E_s^R = E_i + l'E^*$, E_i is the idle energy expenditure, $E_s^T = E_i + l''d^n$, $n = 4$ for long distance transmission, $n = 2$ for short distance transmission, and E^* represents the cost of beam forming approach for energy reduction.

Assuming a first order radio model [62], for a cover with N_s members, the energy expenditure of a cover is the summation of transmission, receiving, and mobility energy cost of all member nodes as follows:

$$E = \sum_{i=1}^{N_s-1} E_{i,s}^T + (kE_s^R + E_{s,B}^T) + E_s^M, \tag{1.4}$$

where the first term is the total energy used to transmit messages from cover members to the head, E_s^R is the energy used by the head node s to receive messages from the member nodes, $E_{s,B}^T$ gives the energy used by the head node s to transmit aggregated messages to the base station, and E_s^M is the moving energy cost used for sensor mobility. In the proposed model, we assume that the working field is fully controlled by the base station. Depending on the transmitted data from sensor nodes after each round, the base station estimates the target speed and its moving direction.

When computing the energy consumption of a sensor node to transmit or receive a message of l bits, we adopt the following formulas for the transmitter energy consumption E^T and receiver energy consumption E^R:

$$E^T = E_e + ld^n \tag{1.5}$$

$$E^R = E_e + lE^* \tag{1.6}$$

where E_e is the idle energy expenditure and d is the distance between the transmitter and receiver. Depending on the distance between the transmitter and receiver, the transmission energy consumption E^T is proportional to different orders of the distance, and it can be modeled with a proper power term n. In the proposed model, we use $n = 4$ for long distance transmission, i.e., transmitting messages from cover head to base station, and $n = 2$ for short distance transmission, i.e., sending messages from a sensor node to its cover head. E^* in Eq. (1.6) represents the cost of beam forming approach to reducing the energy consumption.

The mobility energy cost E_s^M for a sensor s depends on a proposed energy-aware distance $\boldsymbol{\varepsilon}\mathbf{D}$, and $\boldsymbol{\varepsilon}\mathbf{D}$ is computed depending on the Euclidean distance ϱ_d between the starting point x to the end point y, and take into account the environmental hindrances such as trees or buildings. Each hindrance h has weight w which represents its ability to consume the node energy. Let $W = \{w_1, w_2, w_3, \ldots, w_n\}$, so, the $\boldsymbol{\varepsilon}\mathbf{D}$ for a sensor s with sensing field f that contains a set of n hindrances h can be represented by Eq. (1.7). Figure 1.2 depicts sensors mobility in three consecutive rounds to keep covering all target during their movement. Based on targets movement speed and direction, the base station updates sensors locations as shown at Fig. 1.2.

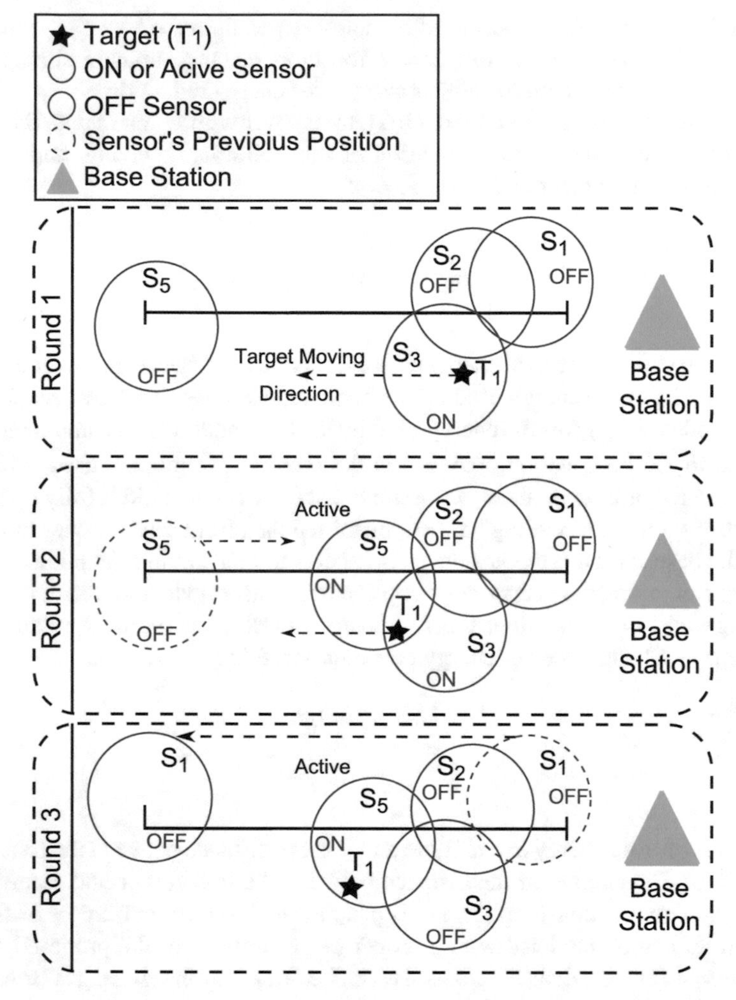

Fig. 1.2 Sensor mobility after each round to keep covering all targets during their movement

$$\boldsymbol{\varepsilon}\mathbf{D}_s = \rho_d\{x, y\} + \sum_{i=0}^{n} h_i w_i \tag{1.7}$$

Using the consumed energy, we can compute the remaining energy of a node s as follows:

$$\tilde{E}_s = E(0) - \sum_{t=1}^{T} (E_s^T(t) + E_s^R(t) + E_s^M(t)) \tag{1.8}$$

where $E(0)$ is the initial energy of the node and t denotes the network lifetime in term of transmission rounds. Note that unless a node serves as the cover head in a round its receiving energy expenditure E^R is zero. In practice, the remaining energy of every node is updated in each round. Each gene in the chromosome represents a sensor node in the field. The value of a gene can be either 1 or 0, where 1 indicates that the corresponding node serves as cover head and 0 indicates a non-head node.

GA generates new chromosomes through crossover and mutation operations and evaluates their fitness. The crossover operation is performed with two randomly selected chromosomes determined by a crossover probability to regulate the operation. When the crossover is excluded, the parent chromosomes are duplicated to the offspring without change. Varying the crossover probability alters the evolution speed of the search process. In practice, the value of crossover is close to 1.

The mutation operation involves altering the value of a randomly selected gene within the chromosome. Similarly, a mutation probability is used to regulate the performance of mutation. Different from the crossover probability, the mutation probability is usually fairly small. Essentially mutation operation could create completely new species, i.e., an arbitrary locus in the fitness landscape. Hence, it is a means to get out of a local optimum.

The fitness function aims to reduce the energy consumption as possible. For that, it evaluates the consumed energy and the expected amount of consumed energy for the proposed sensor cover after each round. Dependently, GA chooses the sensor cover that minimizes the energy exhaustion at a specific data transmission round. Hence, in the model, the GA is not only used to get all possible covers but also, it selects the lowest energy consumption cover at a specific data transmission round. Mathematically, the fitness function of GA consists of the remaining energy $\tilde{E}$, the total expected energy expenditure ΔE, and the distance between the cover head and the base station. Assume that in each round a sensor node transmits a fixed number of bits to the head node, which is then aggregated and relayed to the base station $d(s, B)$. Hence, using Eqs. (1.5) and (1.6) we can compute the energy cost for each node and by aggregating the energy costs of all clusters following Eq. (1.4) the expected energy expenditure ΔE is estimated. The fitness function is hence defined as follows:

$$f = \frac{\tilde{E}}{NE(0)} + \frac{E'}{\Delta E} + \frac{1}{\sum_i d(s_i, B)} \tag{1.9}$$

where E' denote the total energy cost if the messages are transmitted directly from the sensor nodes to the base station. In this fitness function, we normalize the remaining energy and the expected energy expenditure so that they are in the same order.

1.4 Experimental Results and Discussion

In this section, different experiments were conducted to evaluate the performance of the proposed model.

The first experiment was carried out to evaluate the proposed model using three different K-coverage cases. The second experiment was to test the performance regarding the amount of the consumed energy at each sensor node at a specific transmission round. In the third experiment, the aim was to measure the performance of the proposed model compared to the state-of-the-art methods.

1.4.1 Experimental Settings

Table 1.1 lists the network parameters that were used in all experiments. In running GA, we used the population size of 30 for 30 generations. The crossover probability and mutation probability are 0.8 and 0.006, respectively.

In the evaluation of computational time, the experiments were conducted in a PC with Core i5-2400 CPU at 3.1 GHz, 4 GB memory and the system was running Windows 7 and the programs were implemented with MATLAB R2012a.

In the experiments, the period between the start of the network until covering the first target was used as the network lifetime. Each experiment was run ten times, and the average performance of the ten runs was calculated. In each experiment, the nodes of the targets were randomly placed in the field with the condition that all targets are completely covered by the network. In all experiments, there were two different working environments, the dimensions of the first and second environments were 100 m × 100 m and 200 m × 200 m, respectively.

Table 1.1 Network properties

Properties		Values
Number of nodes		100
Initial node energy		0.5 J
Idle state energy		50 nJ/bit
Data aggregation energy		5 nJ/bit
Amplification energy	$d \geq d_0$	10 pJ/bit/m^2
(Cluster head to base-station)	$d < d_0$	0.0013 pJ/bit/m^2
Amplification energy	$d \geq d_1$	$E_{fs}/10 = E_{fs1}$
(Node to cluster head)	$d < d_1$	$E_{mp}/10 = E_{mp1}$
Packet size		400 bits

1.4.2 Proposed Model in a Static Field

In this section, all the experiments were conducted in a static field, i.e., the sensors were fixed.

- In the first experiment, three sub-experiments were carried out to evaluate the proposed model using three different K-coverage cases, where the value of K was 1, 2, and 3. In this experiment, the round time at which the first target becomes uncovered (FTU) and the round time at which the last target become uncovered (LTU) are reported in each experiment. The results of this experiment are summarized in Table 1.2.
  As shown in Table 1.2, as we increased the coverage level, i.e., the value of K, the sensing field increased in the way that makes the sensors consume more energy to transfer the collected data; and hence, the network lifetime decreased. In other words, the network lifetime is inversely proportional to the value of K. The table presents a comparison with three related methods, i.e., FDSSP, FDA, and VP-NL. It is clear that the proposed model improved the performance of the network regarding the network lifetime using different values of K. The improvement that the proposed model achieved was in the range of 26–41.3%.
  Figure 1.3 illustrates a comparison between the proposed model and FDSSP, FDA, and VP-NL methods. The key achievement of the proposed model is the balancing between all sensors regarding the remaining energy by keeping all sensors mostly at the same energy level.
- The second experiment evaluates the performance regarding the amount of the consumed energy at each sensor node at a specific transmission round. The balanced energy exhaustion between all sensors implies the better performance. In this experiment, 100 nodes were randomly placed in a field of 100×100 m with $K = 1$. Ten experiments were performed with randomly placed nodes and targets, and the average energy levels of all nodes at the transmission rounds of 200, 500, and 1000 were visualized.
  Figure 1.4 illustrates the average remaining energy regarding percentage concerning the initial energy of nodes in the field using the proposed model. As shown in Fig. 1.4, the remaining energy is inversely proportional with the number of rounds. Moreover, the network lifetime is extended by avoiding high energy consumption at one node while other nodes keep saving their energy.
- In the third experiment, a comparison was performed regarding the targets covering time. The network lifetime in this experiment is determined by the first target that becomes uncovered. In this experiment, 100 nodes were randomly placed in a 100 m $\times$ 100 m and 200 m $\times$ 200 m environments to cover different numbers of targets. Figure 1.5 depicts the average number of covered targets throughout the entire network lifespan in case of $K = 1$, 2, and 3. In this figure, the nodes were randomly placed in: (a) a 100 m $\times$ 100 m environment to cover 10 targets and (b) a 200 m $\times$ 200 m environment to cover 20 targets. The x-axis is the number of network transmission rounds (in thousands); whereas the y-axis represents the

Table 1.2 Network lifetime for $K = 1$, $K = 2$, and $K = 3$

Coverage	Method	100 m × 100 m		200 m × 200 m	
		FTU	LTU	FTU	LTU
$K = 1$	FDSSP	675	1240	509	914
	FDA	650	1301	630	1000
	VP-NL	751	1514	689	1225
	Proposed model	1050	1847	886	1547
$K = 2$	FDSSP	520	839	414	620
	FDA	487	992	391	785
	VP-NL	586	1053	421	869
	Proposed model	864	1375	627	1148
$K = 3$	FDSSP	312	641	209	300
	FDA	383	787	301	417
	VP-NL	398	803	210	524
	Proposed model	632	1015	402	728

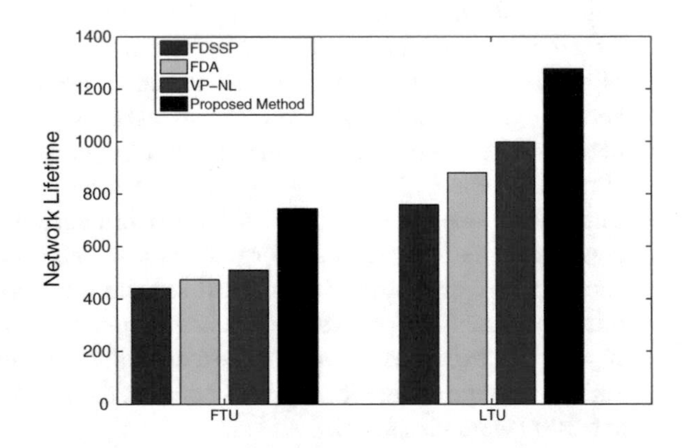

Fig. 1.3 Comparison between the proposed model and FDSSP, FDA, and VP-NL methods in terms of network lifetime

percentage of active nodes. The numbers of nodes deployed in the field in both cases are shown in Table 1.1.

As shown in Fig. 1.5, the network transmission continues, the number of active nodes decreases because more nodes deplete their energy. Figure 1.6 shows an example of sensor placements and covers distributed at the working field using the different cases. It is clear that the proposed model extends the network lifespan by increasing the targets coverage time in all cases.

For further analysis of these experiments, Table 1.3 lists the average and standard deviation of the experimental run time to get the remaining energy of each sensor nodes using the two different cases, i.e., 100 m × 100 m and 200 m × 200 m, that we discussed above, and when the value of K was one.

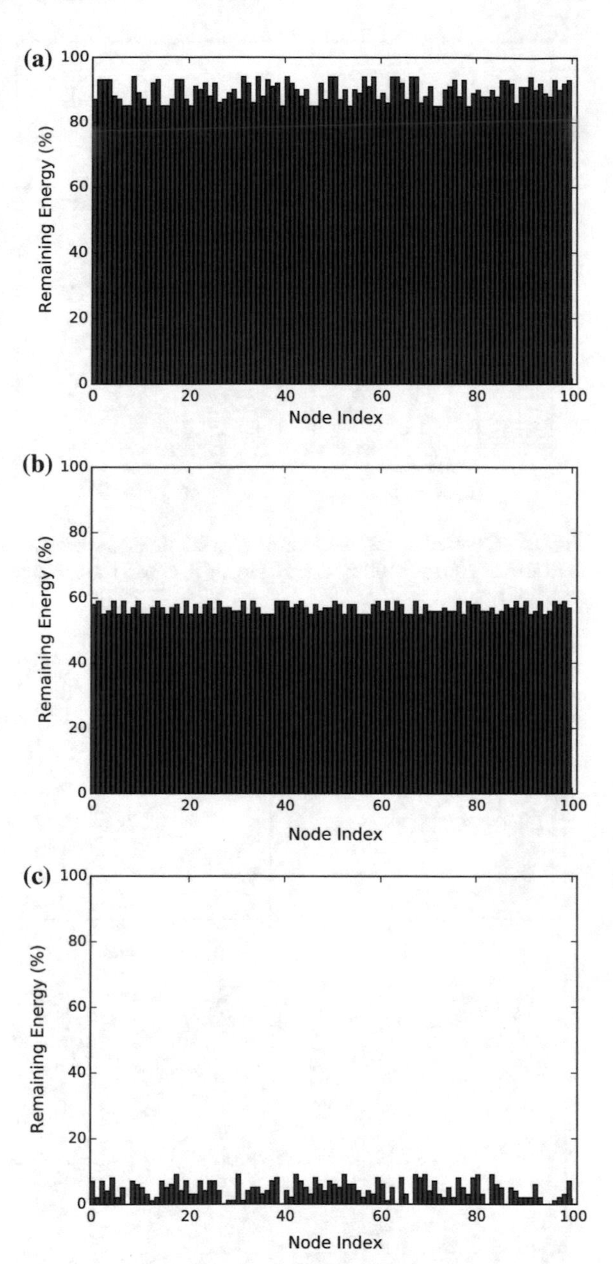

Fig. 1.4 The percentage of the remaining energy of the all network sensor nodes distributed in 100 m × 100 m field. **a** round 200, **b** round 500, and **c** round 1000

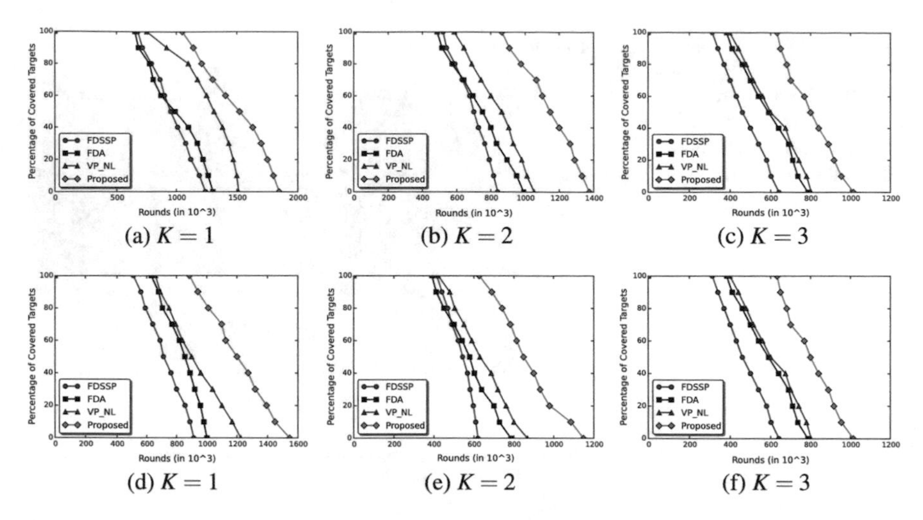

Fig. 1.5 Covered targets in terms of network transmission rounds K. **a**, **b**, and **c** are results of the field size of 100 m × 100 m with 10 targets **d**, **e**, and **f** are results of the field size of 200 m × 200 m with 20 targets

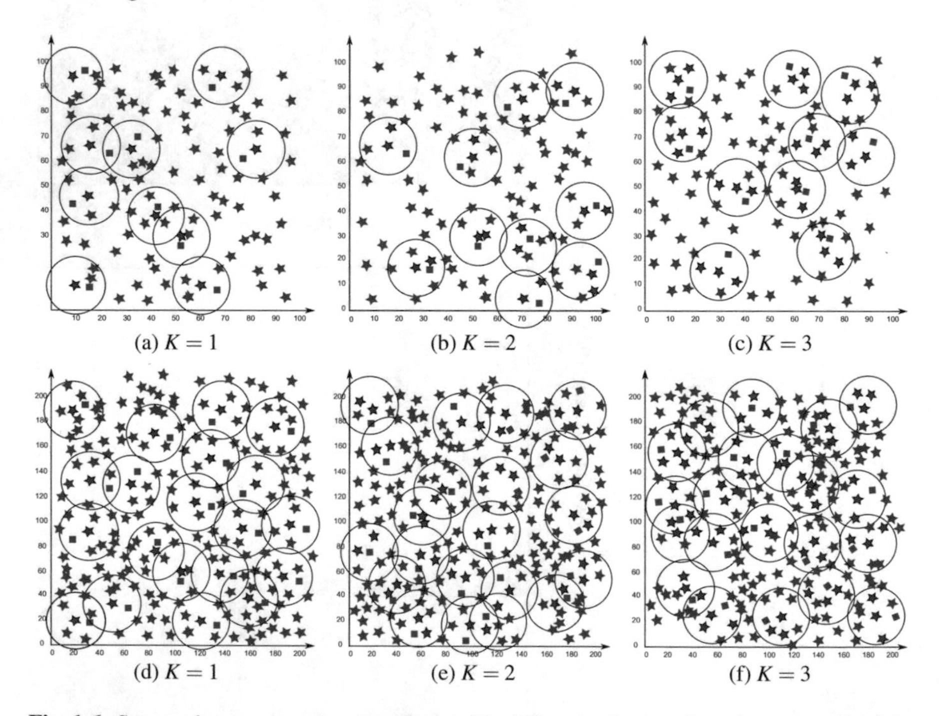

Fig. 1.6 Sensor placements and covers (depicted in different colors). **a**, **b**, and **c** are results of the field size of 100 m × 100 m with 10 targets **d**, **e**, and **f** are results of the field size of 200 m × 200 m with 20 targets

Table 1.3 Average and standard deviation (STD) of the experimental run time to get the remaining energy of each node

Rounds		200	500	800	1100	1300
100 m × 100 m	Mean	0.401	0.332	0.279	0.147	0.332
	STD	0.014	0.037	0.098	0.032	0.074
100 m × 100 m	Mean	0.215	0.321	0.197	0.246	0.207
	STD	0.086	0.049	0.052	0.035	0.044

1.4.3 Proposed Model in a Dynamic Field

In this section, all the experiments were carried out in a dynamic field to measure the effect of sensors mobility on the covering time.

- In the first experiment, the performance of the proposed model was evaluated using three different K-coverage cases, where the value of K was 1, 2, and 3. The results of this experiment are summarized in Fig. 1.7.
  Figure 1.7 depicts the average number of covered targets throughout the entire network lifespan in case of $K = 1, 2,$ and 3. In this figure, the nodes were randomly placed in: (a) a 100 m × 100 m environment to cover 10 targets and (b) a 200 m × 200 m environment to cover 20 targets. As shown in this figure, as the network transmission continues, the number of active nodes decreases as more nodes deplete their energy. Compared to the previous experiments at which all sensors locations are static (see Sect. 1.4.2), Fig. 1.7 illustrates that the network lifetime

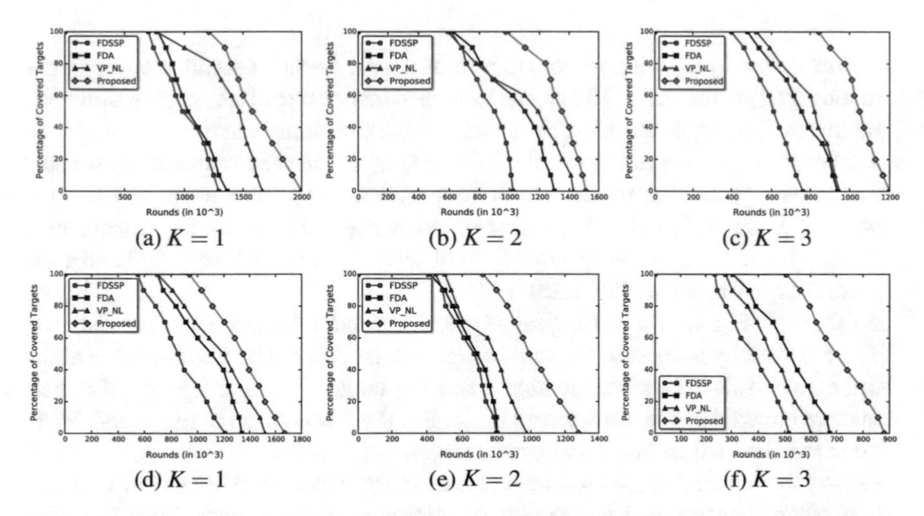

Fig. 1.7 Covered targets in terms of network transmission rounds K with sensors mobility. **a**, **b**, and **c** are results of the field size of 100 m × 100 m with 10 targets **d**, **e**, and **f** are results of the field size of 200 m × 200 m with 20 targets

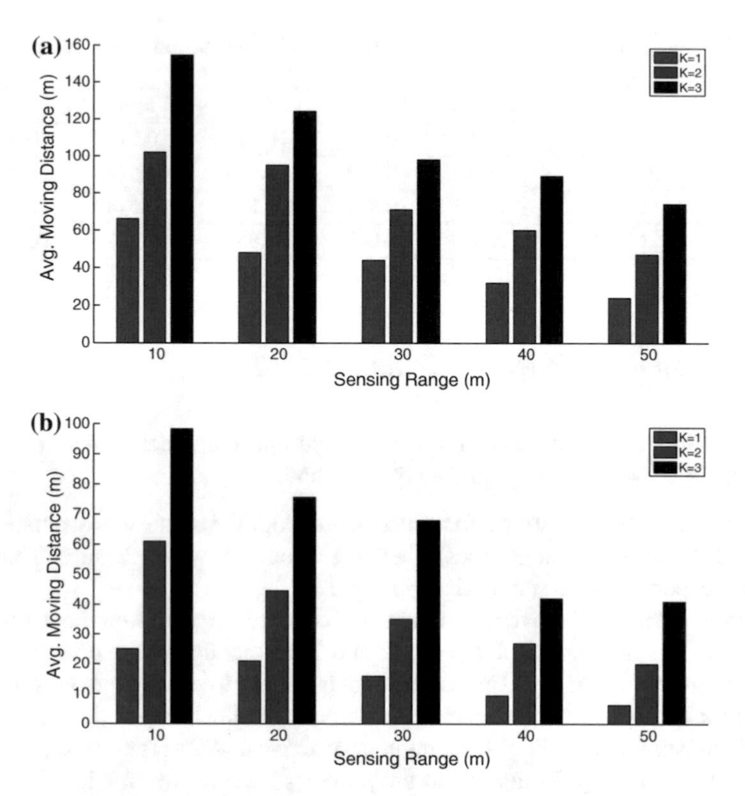

Fig. 1.8 The sensing range versus the average moving distance using 50 sensor nodes distributed at **a** 100 m × 100 m working field **b** 200 m × 200 m working field

was extended by allowing sensors to be mobile. As a result, sensors mobility leads to longer targets covering. Hence, the proposed model extends the network lifespan by increasing targets coverage time in all sensors mobility cases.

- This experiment was run to test the influence of the sensing range of the sensors on the average moving distance. In this experiment, 50 sensors were used, and the value of K was 1, 2, and 3. At the first experiment (see Fig. 1.8a), the sensors were distributed at 100 m × 100 m sensing field while we used 200 m × 200 m for the second experiment (see Fig. 1.8b).
  As shown in Fig. 1.8, the sensing range greatly affects the average moving distance. The reason why is because the same space can be covered by more sensors at the same time. This means that the bigger sensing range, the larger amount of targets that are covered by the same sensor node. So, there is no need to move additional sensors from their locations to cover a target.
  In the same context, a sensor with a bigger sensing range consumes more energy than others. Accordingly, the average remaining energy of a sensor node increases with less sensing range. Figure 1.9 proves this fact using two different cases. At Fig. 1.9a, 50 mobile sensor node are distributed in a field size 100 m × 100 m.

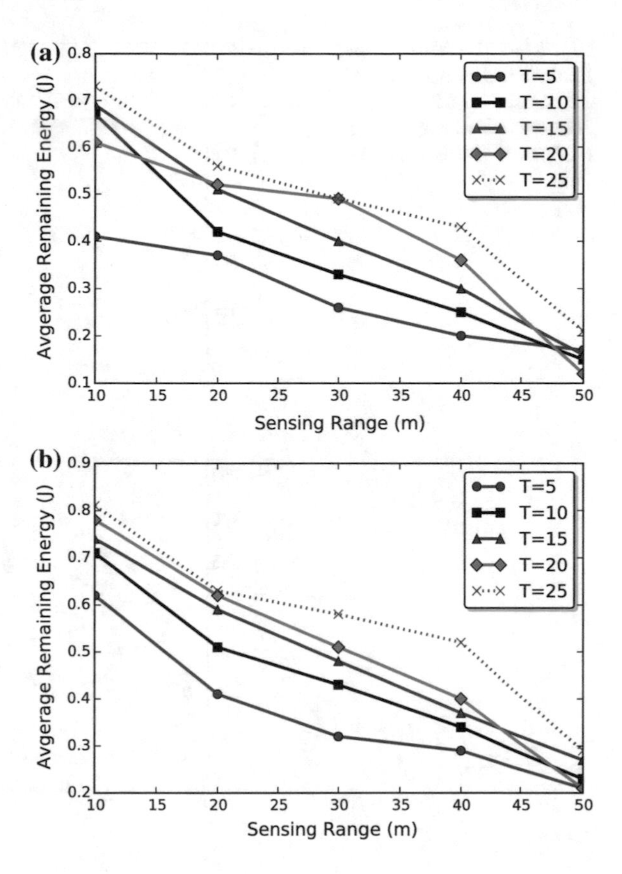

Fig. 1.9 The sensing range versus the remaining energy using a 100 m × 100 m working field with 50 **a** mobile sensor node **b** static sensor node

Applying the same experiment by distributing 50 stationary sensors in a field size 100 m × 100 m leads to the same result as shown in Fig. 1.9b.

- In the most of the conducted experiments, it was remarked that the proposed model avoids the high energy consumption at one node by switching its active status (ON/OFF) according to the current status of its neighbors at each round. This experiment was conducted to test the ability of the proposed model to make a balance between all sensors. The results of this experiment are summarized in Fig. 1.10a and b, where 100 stationary sensor were distributed in a 100 m × 100 m and 200 m × 200 m working fields, respectively.
  As shown in Fig. 1.10, the active time is distributed between all nodes in the way that guarantees the energy consumption balancing. However, in few cases, the active time of one or more sensors is higher than the corresponding active time of the rest sensors that can be occurred when a target or more is covered by a few numbers of sensors. Hence, it is noticed that the active time for the neighbors' sensors is closer.
- The network throughput is completely depending on the amount of transferred data from sensor nodes to the base station. Hence, the longer network lifetime, the higher

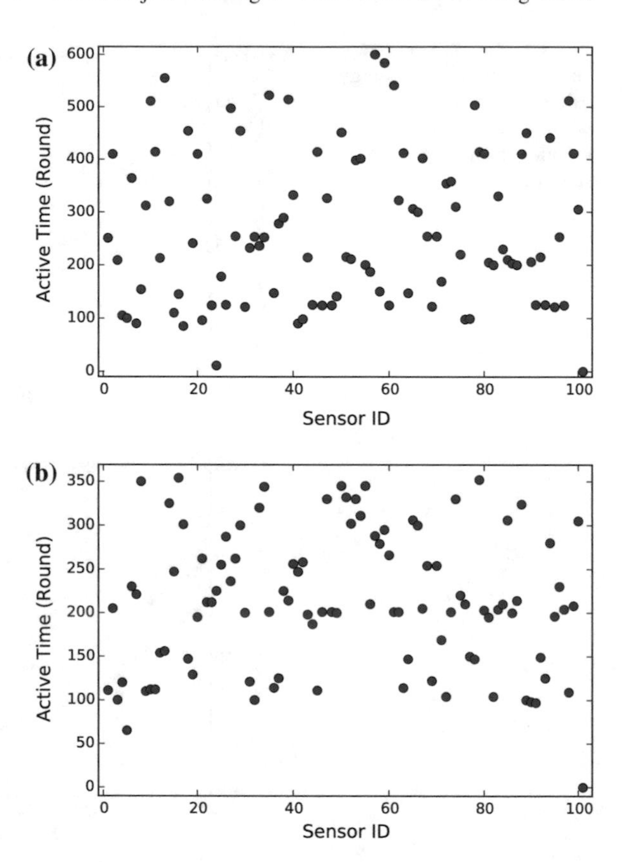

Fig. 1.10 The Active Time for each sensor node using 100 sensors distributed in **a** 100 m × 100 m working field **b** 200 m × 200 m working field

network throughput. As discussed above, the proposed model extends the network lifetime in terms of the first and the last node become unavailable. This experiment was carried out to examine the network throughput of the proposed model. In this experiment, 100 mobile sensor node was distributed in a 100 m × 100 m and 200 m × 200 m working field. The results of this experiment are displayed in Fig. 1.11. As shown in Fig. 1.11a and b it can be remarked that the network throughput inversely proportional with the value of K. In other words, as we increased the coverage level, the network throughput decreased. This is because increasing the coverage level consumes more senor's energy to transfer the collected data; and hence, reduces the network lifetime and consequently network throughput.

- The energy-aware distance is different from node to another. This experiment was carried out in a 100 m × 100 m and 200 m × 200 m working fields, where the number of sensors was 100 in a dynamic environment and $K = 2$. The goal of this experiment is to show how the energy-aware distances are changed across the nodes. The results of this experiment are illustrated in Fig. 1.12.

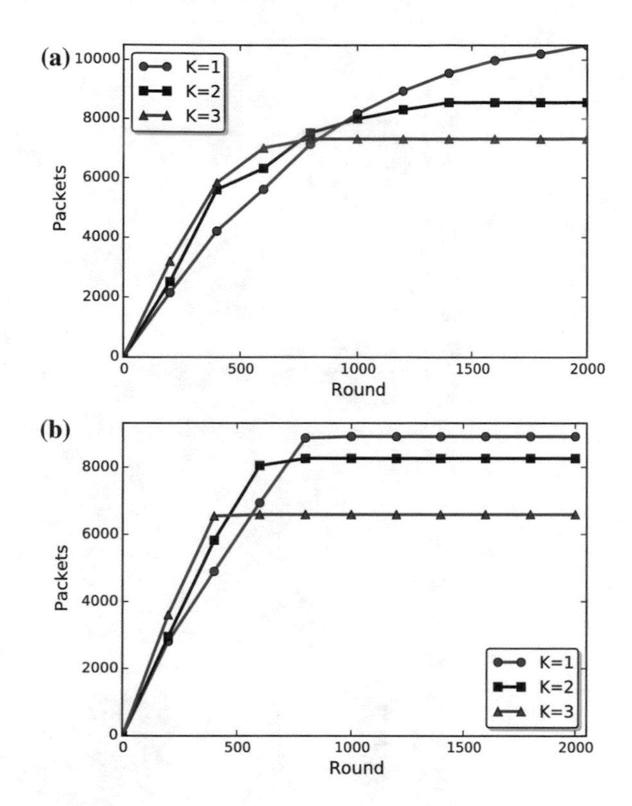

Fig. 1.11 The network throughput using 100 mobile sensor node distributed in **a** 100 m × 100 m working field **b** 200 m × 200 m working field

As shown in Fig. 1.12a and b, this distance is extremely related to the node location. These two figures show the spatial and frequency view of sensor nodes moving to/from different locations throughout the network lifetime. The size of the sphere is proportional with the energy-aware distance a sensor moved to cover a target.

1.5 Summary

This study has two main contributions. Firstly, GA-based cover forming method that creates all possible sensor covers. Secondly, a WSN covers management method that switches between different sensor covers to maximize the network lifetime. The proposed model used the GA to optimize the coverage requirements in WSNs to provide continuous monitoring of specified targets for longest possible time with limited energy resources. Moreover, we allow sensor nodes to move to appropriate positions to collect environmental information. The model is based on the continuous and variable speed movement of mobile sensors to keep all targets under their cover all times. There are three main processes of the GA-based method. First, a binary chromosome is used to encode the sensor nodes within the entire field. Then, GA

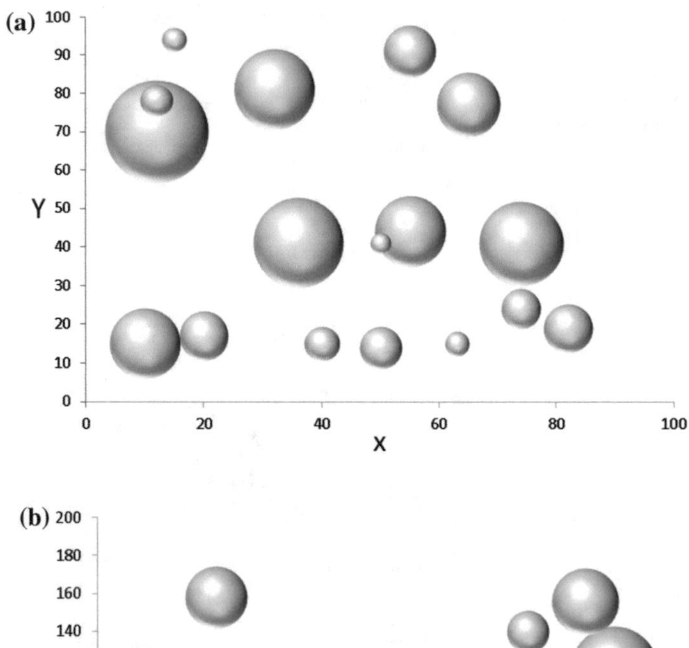

Fig. 1.12 Spatial and frequency view of energy-aware distances for each mobile sensor node using 100 sensor node distributed in 100 m × 100 m working field **b** 200 m × 200 m working field, with $K = 2$

starts working to choose the optimum number of cover heads (represented by 1). Depending on the sensing range of each sensor and the targets positions, covers will be formed. Finally, each chromosome is evaluated to make sure that all targets are covered. A set of experiments in different working environments and a comparison with the most related work are performed.

References

1. Abdelaziz, A., Elhoseny, M., Salama, A. S., & Riad, A. M. (2018). A machine learning model for improving healthcare services on cloud computing environment. *Measurement, 119*, 117–128. https://doi.org/10.1016/j.measurement.2018.01.022.
2. Darwish, A., Hassanien, A. E., Elhoseny, M., Sangaiah, A. K., & Muhammad, K. (2017). The impact of the hybrid platform of internet of things and cloud computing on healthcare systems: Opportunities, challenges, and open problems. *Journal of Ambient Intelligence and Humanized Computing*. Springer. https://doi.org/10.1007/s12652-017-0659-1.
3. Elhoseny, M., Abdelaziz, A., Salama, A. S., Riad, A. M., Muhammad, K., & Sangaiah, A. K. (2018). A hybrid model of internet of things and cloud computing to manage big data in health services applications. *Future Generation Computer Systems*. Elsevier. (in Press, Accepted March 2018).
4. Elhoseny, M., Nabil, A., Hassanien, A. E., & Oliva, D. (2018). Hybrid rough neural network model for signature recognition. In A. Hassanien, & D. Oliva (Eds.), *Advances in soft computing and machine learning in image processing*. Studies in Computational Intelligence, Vol. 730. Cham: Springer. https://doi.org/10.1007/978-3-319-63754-9_14.
5. Elhoseny, M., Tharwat, A., Farouk, A., & Hassanien, A. E. (2017b). K-coverage model based on genetic algorithm to extend WSN lifetime. *IEEE Sensors Letters, 1*(4), 1–4.
6. Elsayed, W., Elhoseny, M., Riad, A., & Hassanien, A. E. (2017). Autonomic self-healing approach to eliminate hardware faults in wireless sensor networks. In *International conference on advanced intelligent systems and informatics*, pp. 151–160. Springer.
7. Sajjad, M., Nasir, M., Muhammad, K., Khan, S., Jan, Z., Sangaiah, A. K., Elhoseny, M., & Baik, S. W. (2018). Raspberry Pi assisted face recognition framework for enhanced law-enforcement services in smart cities. *Future Generation Computer Systems*. Elsevier. https://doi.org/10.1016/j.future.2017.11.013.
8. Shehab, A., Elhoseny, M., El Aziz, M. A., & Hassanien, A. E. (2018) Efficient schemes for playout latency reduction in P2P-VoD systems. In A. Hassanien, & D. Oliva (Eds.), *Advances in soft computing and machine learning in image processing*. Studies in Computational Intelligence, Vol. 730. Cham: Springer. https://doi.org/10.1007/978-3-319-63754-9_22.
9. Tuna, G., Gungor, V., Gulez, K., Hancke, G., & Gungor, V. (2013). Energy harvesting techniques for industrial wireless sensor networks. In G. P. Hancke & V. C. Gungor (Eds.),*Industrial wireless sensor networks: applications, protocols, standards, and products*, pp. 119–136.
10. Yuan, X., Li, D., Mohapatra, D., & Elhoseny, M. (2017). Automatic removal of complex shadows from indoor videos using transfer learning and dynamic thresholding. *Computers and Electrical Engineering*. https://doi.org/10.1016/j.compeleceng.2017.12.026. (in Press).
11. Das, A., & Bruhadeshwar, B. (2013). A biometric-based user authentication scheme for heterogeneous wireless sensor networks. In *27th international conference on advanced information networking and applications workshops (WAINA)*, (pp. 291–296). IEEE.
12. Elhoseny, M., Hosny, A., Hassanien, A. E., Muhammad, K., & Sangaiah, A. K. (2017). Secure automated forensic investigation for sustainable critical infrastructures compliant with green computing requirements. *IEEE Transactions on Sustainable Computing*, *PP*(99). https://doi.org/10.1109/TSUSC.2017.2782737.
13. Elhoseny, M., Tharwat, A., & Hassanien, A. E. (2017c). Bezier curve based path planning in a dynamic field using modified genetic algorithm. *Journal of Computational Science*. https://doi.org/10.1016/j.jocs.2017.08.004.
14. Ferentinos, K., Katsoulas, N., Tzounis, A., Bartzanas, T., & Kittas, C. (2017). Wireless sensor networks for greenhouse climate and plant condition assessment. *Biosystems Engineering*, *153*, 70–81.

15. Abdelaziz, A., Elhoseny, M., Salama, A. S., Riad, A. M., & Hassanien, A. E. (2018). Intelligent algorithms for optimal selection of virtual machine in cloud environment, towards enhance healthcare services. In A. Hassanien, K. Shaalan, T. Gaber, & M. Tolba (Eds.), *2017 Proceedings of the international conference on advanced intelligent systems and informatics, AISI 2017*. Advances in Intelligent Systems and Computing, Vol. 639. Cham: Springer. https://doi.org/10.1007/978-3-319-64861-3_27.
16. Elsayed, W., Elhoseny, M., Riad, A. M., & Hassanien, A. E. (2018). Autonomic self-healing approach to eliminate hardware faults in wireless sensor networks. In A. Hassanien, A. Shaalan, T. Gaber, & M. Tolba (Eds.), *2017 Proceedings of the international conference on advanced intelligent systems and informatics, AISI 2017*. Advances in Intelligent Systems and Computing, Vol. 639. Cham: Springer. https://doi.org/10.1007/978-3-319-64861-3_14
17. Hackmann, G., Guo, W., Yan, G., Sun, Z., Lu, C., & Dyke, S. (2014). Cyber-physical codesign of distributed structural health monitoring with wireless sensor networks. *IEEE Transactions on Parallel and Distributed Systems*, *25*(1), 63–72.
18. Hassanien, A. E., Tharwat, A., & Own, H. S. (2017). Computational model for Vitamin D deficiency using hair mineral analysis. *Computational Biology and Chemistry*.
19. Shahin, M. K., Tharwat, A., Gaber, T., & Hassanien, A. E. (2017). A wheelchair control system using human-machine interaction: Single-modal and multimodal approaches. *Journal of Intelligent Systems*.
20. Shehab, A., Ismail, A., Osman, L., Elhoseny, M., & El-Henawy, I. M. (2018). Quantified self using IoT wearable devices. In A. Hassanien, K. Shaalan, T. Gaber, & M. Tolba (Eds.), *2017 Proceedings of the international conference on advanced intelligent systems and informatics, AISI 2017*. Advances in Intelligent Systems and Computing, Vol. 639. Cham: Springer. https://doi.org/10.1007/978-3-319-64861-3_77.
21. Tharwat, A., Moemen, Y. S., & Hassanien, A. E. (2016b). A predictive model for toxicity effects assessment of biotransformed hepatic drugs using iterative sampling method. *Scientific Reports*, *6*, 38660.
22. Tharwat, A., Moemen, Y. S., & Hassanien, A. E. (2017). Classification of toxicity effects of biotransformed hepatic drugs using whale optimized support vector machines. *Journal of Biomedical Informatics*, *68*, 132–149.
23. Abdeldaim, A. M., Sahlol, A. T., Elhoseny, M., & Hassanien, A. E. (2018). Computer-aided acute lymphoblastic leukemia diagnosis system based on image analysis. In A. Hassanien, & D. Oliva (Eds.), *Advances in soft computing and machine learning in image processing*. Studies in Computational Intelligence, Vol. 730. Cham: Springer. https://doi.org/10.1007/978-3-319-63754-9.
24. Elhoseny, H., Elhoseny, M., Riad, A. M., & Hassanien, A. E. (2018). A framework for big data analysis in smart cities. In A. Hassanien, M. Tolba, M. Elhoseny, & M. Mostafa (Eds.) *AMLTA 2008 the international conference on advanced machine learning technologies and applications* (AMLTA2018). Advances in Intelligent Systems and Computing, Vol. 723. Cham: Springer. https://doi.org/10.1007/978-3-319-74690-6_40.
25. Elhoseny, M., Shehab, A., & Osman, L. (2018). An empirical analysis of user behavior for P2P IPTV workloads. In A. Hassanien, M. Tolba, M. Elhoseny, & M. Mostafa (Eds.), *AMLTA 2008 the international conference on advanced machine learning technologies and applications* (AMLTA2018). Advances in Intelligent Systems and Computing, Vol. 723. Cham: Springer. https://doi.org/10.1007/978-3-319-74690-6_25.
26. Gaber, T., Tharwat, A., Hassanien, A. E., & Snasel, V. (2016). Biometric cattle identification approach based on webers local descriptor and adaboost classifier. *Computers and Electronics in Agriculture*, *122*, 55–66.
27. Srbinovska, M., Gavrovski, C., Dimcev, V., Krkoleva, A., & Borozan, V. (2015). Environmental parameters monitoring in precision agriculture using wireless sensor networks. *Journal of Cleaner Production*, *88*, 297–307.
28. Tharwat, A., Gaber, T., & Hassanien, A. E. (2016a). One-dimensional vs. two-dimensional based features: Plant identification approach. *Journal of Applied Logic*.

29. Wang, M. M., Qu, Z. G., & Elhoseny, M. (2017). Quantum secret sharing in noisy environment. In X. Sun, H. C. Chao, X. You, & E. Bertino (Eds.), *Cloud computing and security, ICCCS 2017*. Lecture Notes in Computer Science, Vol. 10603. Cham: Springer. https://doi.org/10.1007/978-3-319-68542-7_9.
30. Cerulli, R., Donato, R. D., & Raiconi, A. (2012). Exact and heuristic methods to maximize network lifetime in wireless sensor networks with adjustable sensing ranges. *European Journal of Operational Research*, *220*(1), 58–66.
31. Elhoseny, M., Yuan, X., El-Minir, H. K., Riad, A. (2014). Extending self-organizing network availability using genetic algorithm. In *International conference on computing, communication and networking technologies (ICCCNT)*, (pp. pp. 1–6). IEEE.
32. Elhoseny, M., Elminir, H., Riad, A., & Yuan, X. (2016a). A secure data routing schema for wsn using elliptic curve cryptography and homomorphic encryption. *Journal of King Saud University-Computer and Information Sciences*, *28*(3), 262–275.
33. Elhoseny, M., Yuan, X., El-Minir, H. K., & Riad, A. M. (2016b). An energy efficient encryption method for secure dynamic WSN. *Security and Communication Networks*, *9*(13), 2024–2031.
34. Yang, Q., He, S., Li, J., Chen, J., & Sun, Y. (2015). Energy-efficient probabilistic area coverage in wireless sensor networks. *IEEE Transactions on Vehicular Technology*, *64*(1), 367–377.
35. Hosseinabadi, A. A. R., Vahidi, J., Saemi, B., Sangaiah, A. K., & Elhoseny, M. (2008). Extended genetic algorithm for solving open-shop scheduling problem. *Soft Computing*. https://doi.org/10.1007/s00500-018-3177-y.
36. Elhoseny, M., Yuan, X., ElMinir, H. K., & Riad, A. M. (2016). An energy efficient encryption method for secure dynamic WSN. *Security and Communication Networks*, *9*(13): 2024–2031. https://doi.org/10.1002/sec.1459.
37. Katsuma, R., Murata, Y., Shibata, N., Yasumoto, K., & Ito, M. (2009). Extending k-coverage lifetime of wireless sensor networks using mobile sensor nodes. In *IEEE international conference on wireless and mobile computing, networking and communications*, (pp. 48–54). IEEE.
38. Liu, Z. (2007). Maximizing network lifetime for target coverage problem in heterogeneous wireless sensor networks. In *International conference on mobile Ad-Hoc and sensor networks*, pp. 457–468. Springer.
39. Lu, Z., Li, W., & Pan, M. (2015). Maximum lifetime scheduling for target coverage and data collection in wireless sensor networks. *IEEE Transactions on Vehicular Technology*, *64*(2), 714–727.
40. Elhoseny, M., Tharwat, A., Yuan, X., & Hassanien, A. E. (2018). Optimizing K-coverage of mobile WSNs. *Expert Systems with Applications*, *92*, 142–153. https://doi.org/10.1016/j.eswa.2017.09.008.
41. Wan, X., Wu, J., & Shen, X. (2015). Maximal lifetime scheduling for roadside sensor networks with survivability. *IEEE Transactions on Vehicular Technology*, *64*(11), 5300–5313.
42. Tang, J., Zhu, B., Zhang, L., & Hincapie, R. (2011). Wakeup scheduling in roadside directional sensor networks. In *Global telecommunications conference (GLOBECOM 2011)*, pp. 1–6. IEEE.
43. Mnasri, S., Thaljaoui, A., Nasri, N., & Val, T. (2015). A genetic algorithm-based approach to optimize the coverage and the localization in the wireless audio-sensors networks. In *International symposium on networks, computers and communications (ISNCC)*, pp. 1–6. IEEE.
44. Yang, Q., & Gündüz, D. (2015). Variable-power scheduling for perpetual target coverage in energy harvesting wireless sensor networks. In *International symposium on wireless communication systems (ISWCS)*, pp. 281–285. IEEE.
45. Han, G., Jiang, J., Zhang, C., Duong, T., Guizani, M., & Karagiannidis, G. (2016). A survey on mobile anchor node assisted localization in wireless sensor networks. *IEEE Communications Surveys & Tutorials*, *18*(3), 2220–2243.
46. Wang, R., Xu, B., Wei, R., Gu, H., & Chen, J. (2010). Design and implementation of an intelligent environmental monitoring system for animal house based on wireless sensor net (WSN)[j]. *Jiangsu Journal of Agricultural Sciences*, *3*, 024.
47. Hwang, J., Shin, C., & Yoe, H. (2010). Study on an agricultural environment monitoring server system using wireless sensor networks. *Sensors*, *10*(12), 11189–11211.

48. Elhoseny, M., Elleithy, K., Elminir, H., Yuan, X., & Riad, A. (2015). Dynamic clustering of heterogeneous wireless sensor networks using a genetic algorithm towards balancing energy exhaustion. *International Journal of Scientific & Engineering Research*, *6*(8), 1243–1252.
49. Fadel, E., Gungor, V., Nassef, L., Akkari, N., Malik, M., Almasri, S., et al. (2015). A survey on wireless sensor networks for smart grid. *Computer Communications*, *71*, 22–33.
50. Rawat, P., Singh, K., Chaouchi, H., & Bonnin, J. (2014). Wireless sensor networks: a survey on recent developments and potential synergies. *The Journal of supercomputing*, *68*(1), 1–48.
51. Elhoseny, M., Elkhateb, A., Sahlol, A., & Hassanien, A. E. (2018). Multimodal biometric personal identification and verification. In A. Hassanien, & D. Oliva (Eds.), *Advances in soft computing and machine learning in image processing*. Studies in Computational Intelligence, Vol. 730. Cham: Springer. https://doi.org/10.1007/978-3-319-63754-9_12.
52. Elhoseny, M., Essa, E., Elkhateb, A., Hassanien, A. E., & Hamad, A. (2018). Cascade multimodal biometric system using fingerprint and Iris patterns. In A. Hassanien, K. Shaalan, T. Gaber, & M. Tolba (Eds.) *2017 Proceedings of the international conference on advanced intelligent systems and informatics, AISI 2017*. Advances in Intelligent Systems and Computing, Vol. 639. Cham: Springer. https://doi.org/10.1007/978-3-319-64861-3_55.
53. Elhoseny, M., Ramírez-Gonz, G., & Farouk, A. (2018). Secure medical data transmission model for IoT-based healthcare systems. *IEEE AccessPP*(99). https://doi.org/10.1109/ACCESS.2018.2817615.
54. Farouk, A., Batle, J., Elhoseny, M., Naseri, M., Lone, M., Fedorov, A., Alkhambashi, M., Ahmed, S. H., & Abdel-Aty, M. (2018). Robust general N user authentication scheme in a centralized quantum communication network via generalized GHZ states. *Frontiers of Physics13* 130306. Springer. https://doi.org/10.1007/s11467-017-0717-3
55. Shehab, A., Elhoseny, M., Muhammad, K., Sangaiah, A. K., Yang, P., Huang, H., & Hou, G. (2018). Secure and robust fragile watermarking scheme for medical images. *IEEE Access*, *6*(1), pp. 10269–10278. https://doi.org/10.1109/ACCESS.2018.2799240.
56. Elhoseny, M., Shehab, A., & Yuan, X. (2017). Optimizing robot path in dynamic environments using genetic algorithm and Bezier curve. *Journal of Intelligent & Fuzzy Systems*, *334*, 2305–2316. IOS-Press. https://doi.org/10.3233/JIFS-17348.
57. Metawaa, N., Kabir Hassana, M., & Elhoseny, M. (2017). Genetic algorithm based model for optimizing bank lending decisions. *Expert Systems with Applications*, *80*, 75–82. https://doi.org/10.1016/j.eswa.2017.03.02.
58. Ebrahimian, N., Sheramin, G., Navin, A., & Foruzandeh, Z. (2010). A novel approach for efficient k-coverage in wireless sensor networks by using genetic algorithm. In *International conference on computational intelligence and communication networks (CICN)*, (pp. 372–376). IEEE.
59. Elhoseny, M., Farouk, A., Zhou, N., Wang, M., Abdalla, S., & Batle, J. (2017a). Dynamic multihop clustering in a wireless sensor network: Performance improvement. *Wireless Personal Communications*, 1–21.
60. Elhoseny, M., Yuan, X., Yu, Z., Mao, C., El-Minir, H., & Riad, A. (2015). Balancing energy consumption in heterogeneous wireless sensor networks using genetic algorithm. *IEEE Communications Letters*, *19*(12), 2194–2197.
61. Shieh, C., Sai, V., Lin, Y., Lee, T., Nguyen, T., & Le, Q. (2016). Improved node localization for WSN using heuristic optimization approaches. In *International conference on networking and network applications (NaNA)*, pp. 95–98. IEEE.
62. Yuan, X., Elhoseny, M., El-Minir, H., & Riad, A. (2017). A genetic algorithm-based, dynamic clustering method towards improved WSN longevity. *Journal of Network and Systems Management*, *25*(1), 21–46.
63. Berman, P., Calinescu, G., Shah, C., & Zelikovsky, A. (2004). Power efficient monitoring management in sensor networks. In *Proceedings of the wireless communications and networking conference (WCNC)*, (Vol. 4, pp. 2329–2334). IEEE.
64. Cardei, M., & Du, D. (2005). Improving wireless sensor network lifetime through power aware organization. *Wireless Networks*, *11*(3), 333–340.

65. Slijepcevic, S., & Potkonjak, M. (2001). Power efficient organization of wireless sensor networks. In *IEEE international conference on communications (ICC)*, *2*, 472–476. IEEE.
66. Cardei, M., Thai, M., Li, Y., & Wu, W. (2005a). Energy-efficient target coverage in wireless sensor networks. In *Proceedings IEEE 24th annual joint conference of the IEEE computer and communications societies*, (Vol. 3, pp. 1976–1984). IEEE.
67. Cardei, M., Wu, J., Lu, M., & Pervaiz, M. (2005b). Maximum network lifetime in wireless sensor networks with adjustable sensing ranges. In *Proceedings IEEE international conference on wireless and mobile computing, networking and communications*, (WiMob'2005), (Vol. 3, pp. 438–445). IEEE.
68. Lu, M., Wu, J., Cardei, M., & Li, M. (2005). Energy-efficient connected coverage of discrete targets in wireless sensor networks. In: *Networking and mobile computing*, pp. 43–52. Springer.
69. Liu, H., Wan, P., Yi, C., Jia, X., Makki, S., & Pissinou, N. (2005). Maximal lifetime scheduling in sensor surveillance networks. In *Proceedings IEEE 24th annual joint conference of the IEEE computer and communications societies*, (Vol. 4, pp. 2482–2491). IEEE.
70. Liu, H., Wan, P., & Jia, X. (2006). Maximal lifetime scheduling for k to 1 sensor-target surveillance networks. *Computer Networks*, *50*(15), 2839–2854.
71. Wang, X., Ma, J., Wang, S., & Bi, D. (2007). Distributed particle swarm optimization and simulated annealing for energy-efficient coverage in wireless sensor networks. *Sensors*, *7*(5), 628–648.
72. Mini, S., Udgata, S., & Sabat, S. (2011). Artificial bee colony based sensor deployment algorithm for target coverage problem in 3-D terrain. In *International conference on distributed computing and internet technology*, pp. 313–324. Springer.
73. Huang, Y., & Li, K. (2013). Coverage optimization of wireless sensor networks based on artificial fish swarm algorithm. *Jisuanji Yingyong Yanjiu*, *30*(2), 554–556.
74. Wang, G., Guo, L., Duan, H., Liu, L., & Wang, H. (2012). Dynamic deployment of wireless sensor networks by biogeography based optimization algorithm. *Journal of Sensor and Actuator Networks*, *1*(2), 86–96.
75. Maleki, I., Khaze, S., Tabrizi, M., & Bagherinia, A. (2013). A new approach for area coverage problem in wireless sensor networks with hybrid particle swarm optimization and differential evolution algorithms. *International Journal of Mobile Network Communications & Telematics (IJMNCT)*, *3*.
76. Elhoseny, M., Elminir, H., Riad, A. M., & Yuan, X. I. (2014). Recent advances of secure clustering protocols in wireless sensor networks. *International Journal of Computer Networks and Communications Security*, *2*(11), 400–413.
77. Elsayed, W., Elhoseny, M., Sabbeh, S., & Riad, A. (2007). Self-maintenance model for wireless sensor networks. *Computers and Electrical Engineering*. https://doi.org/10.1016/j.compeleceng.2017.12.022. (in Press).
78. Elhoseny, M., Farouk, A., Batle, J., Shehab, A., & Hassanien, A. E. (2017). Secure image processing and transmission schema in cluster-based wireless sensor network. In *Handbook of research on machine learning innovations and trends*, (Chapter 45, pp. 1022–1040), IGI Global. https://doi.org/10.4018/978-1-5225-2229-4.ch045.
79. Batle, J., Naseri, M., Ghoranneviss, M., Farouk, A., Alkhambashi, M., & Elhoseny, M. (2017). Shareability of correlations in multiqubit states: Optimization of nonlocal monogamy inequalities. *Physical Review A*, *95*(3), 032123. https://doi.org/10.1103/PhysRevA.95.032123.
80. El Aziz, M. A., Hemdan, A. M., Ewees, A. A., Elhoseny, M., Shehab, A., Hassanien, A. E., & Xiong, S. (2017). Prediction of biochar yield using adaptive neuro-fuzzy inference system with particle swarm optimization. In *2017 IEEE PES PowerAfrica conference*, (pp. 115–120), June 27–30, 2017, Accra-Ghana: IEEE. https://doi.org/10.1109/PowerAfrica.2017.7991209.
81. Ewees, A. A., Aziz, M. A. E., & Elhoseny, M. (2007). Social-spider optimization algorithm for improving ANFIS to predict biochar yield. In *2017 8th international conference on computing, communication and networking technologies (8ICCCNT)*, July 3–5 2007. Delhi-India: IEEE.
82. Metawa, N., Elhoseny, M. Hassan, M. K., & Hassanien, A. E. (2006). Loan portfolio optimization using genetic algorithm: A case of credit constraints. In *2016 Proceedings of 12th international computer engineering conference (ICENCO)*, pp. 59–64. IEEE. https://doi.org/10.1109/ICENCO.2016.7856446.

83. Rizk-Allah, R. M., Hassanien, A. E., & Elhoseny, M. (2018). A multi-objective transportation model under neutrosophic environment. *Computers and Electrical Engineering*. Elsevier. https://doi.org/10.1016/j.compeleceng.2018.02.024. (in Press).
84. Tharwat, A., Elhoseny, M., Hassanien, A. E., Gabel, T., & Kumar, A. (2018). Intelligent Bezir curve-based path planning model using chaotic particle swarm optimization algorithm. *Cluster Computing*, 1–22. Springer. https://doi.org/10.1007/s10586-018-2360-3.
85. Tharwat, A., Mahdi, H., Elhoseny, M., & Hassanien, A. E. (2008). Recognizing human activity in mobile crowdsensing environment using optimized kNN algorithm. *Expert Systems With Applications*. https://doi.org/10.1016/j.eswa.2018.04.017. Accessed 12 April 2018.

Chapter 2
Expand Mobile WSN Coverage in Harsh Environments

Abstract Ideally, the lifetime of a homogeneous WSN is maximized when the remaining energy of nodes in the network remains the same. However, most of WSN applications in harsh and complex environments require a kind of nodes heterogeneity, i.e., node mobility; to extend the network coverage and lifetime. In homogeneous WSN, clustering protocols assumed that all the sensor nodes are supplied with the same characteristics, i.e., initial energy. However, placing few heterogeneous nodes in WSN, such as nodes with more computing powers, is an effective way to increase network lifetime and reliability. In this chapter, we propose a sensor clustering method for dynamically organizing heterogeneous WSN using Genetic Algorithm. Moreover, we propose a set of key heterogeneity factors that enhance the performance of WSNs in harsh environments.

2.1 Introduction

The clustering model has been used in Wireless Sensor Networks (WSNs) to improve its performance regarding network availability [1–4]. Although the great works in the process of forming clusters, the dynamic nature of WSN and numerous possible cluster configurations make searching for an optimal network structure [5, 6], a complicated defy especially in case of heterogeneous WSN model [8–10]. This model is an adapted model of homogeneous clustering model, i.e., LEACH [11] through placing few heterogeneous nodes in network [12–14], such as nodes with more computing powers. Allowing the sensors to be mobile in the heterogeneous model increases the number of WSN applications [15–24] compared with static sensors, i.e., tracking animal movements applications [25]. In homogeneous WSN, clustering protocols assumed that all the sensor nodes are supplied with the same characteristics, i.e., initial energy. LEACH is a popular example of homogeneous clustering model. However, placing few heterogeneous nodes in WSN, such as nodes with more computing powers, is an effective way to increase network lifetime and reliability. In a heterogeneous WSN, in addition to the network structuring factors, e.g., distance to the base-station, and distance among nodes, factors such as initial energy, data processing capability, ability to serve as cluster head greatly influence

M. Elhoseny and A. E. Hassanien, *Dynamic Wireless Sensor Networks*, Studies in Systems, Decision and Control 165, https://doi.org/10.1007/978-3-319-92807-4_2

the network lifespan [26–28]. Moreover, many security factors [29–33] should be taken into account, as we will discuss in the coming chapters.

To extend the network lifetime in a heterogeneous network, methods have been proposed that account for one or more of these factors.The factors selection depends on the normal of the application and its environment [34–37]. For each application, many algorithms are proposed to optimize [38–42] these factors selection. Stable Election Protocol (SEP) [14] used weighted probabilities to elect cluster heads depending on the remaining energy in sensor nodes. Also, Developed Distributed Energy-Efficient Clustering (DDEEC) [12] method improved upon SEP by categorizing sensor nodes based on their energy level. The nodes with higher energy were the "advanced nodes", and the cluster head was selected from these group of nodes. Threshold Sensitive Stable Election Protocol (TSEP) [13] extended SEP method by grouping sensor nodes into three energy levels, and the cluster heads were selected based on thresholds. Similarly, Energy Efficient Heterogeneous Clustered scheme (EEHC) [43] and Efficient Three Level Energy algorithm (ETLE) [44] selected cluster heads based on the probability proportional to the residual energy. In Hybrid Energy Efficient Reactive protocol (HEER) [45], the cluster head selection is based on the ratio of the residual energy of nodes and the average energy of the network. Both of Energy efficient heterogeneous clustered scheme (EEHC) [43] and Efficient Three Level Energy algorithm (ETLE) [44] assume three levels of heterogeneity and nodes are randomly distributed and are stationary. In EEHC, the cluster heads were selected based on weighted election probabilities of each node according to the residual energy. While in ETLE, each node chose a random number between 0 and 1. If the value of this random number was less than a threshold value, i.e., T, the node will be selected to serve as a cluster head. In Hybrid Energy Efficient Reactive protocol (HEER) [45], the CH selection is based on the ratio of the residual energy of nodes and the average energy of the network. All of these methods were proposed for WSN with initial energy heterogeneity only.

Ideally, the lifetime of a homogeneous WSN is maximized when the remaining energy of nodes in the network remains the same [46–49]; that is, no single node completely depletes its energy before the others. This is, however, difficult to achieve in a real-world cluster-based network due to different roles of sensor nodes and various signal transmission distance. The nodes serving as cluster head consume additional energy to fulfill tasks such as receiving messages from member nodes and relaying the aggregated messages to the base station. Balancing node energy consumption and extending the overall network lifespan are non-trivial given many factors that could affect the energy expenditure of each node [50, 51].

Searching for a balance among many factors is non-trivial. Optimization methods, such as Genetic Algorithm (GA) [52–55], have been employed in the routing protocol of WSN [56–61]. Genetic algorithm uses random search to suggest the best appropriate design. We use this algorithm to obtain the most efficient clustering structure. The reason for choosing GA is its convergence and its flexibility in solving multi-objective optimization problems like dynamic clustering of WSN. When GA is used, a key objective is to define an appropriate fitness function that encodes the network structure. However, most of the GA-based work was developed for the homogeneous model, i.e., HCR [56, 62], while the remaining was concerned with heterogeneous WSN in which the difference between sensors in the initial energy is the dominate factor of heterogeneity. The Evolutionary Based Clustered Routing Protocol (ERP) [57] overcame the limitations of clustering-algorithm-based GAs by uniting the clustering aspects of cohesion and separation error and proposed a new fitness function based on these two aspects.

Although most of the current research concentrated on energy as the only heterogeneity factor, many types of heterogeneous resources, e.g., communication capability, data processing power, and efficiency, were introduced to WSN for improved performance. Providing sensor node with more processing capabilities aims to prevent it from exhausting its energy quickly in case of acting as a cluster head. On the other hand, preventing some nodes to serve as a cluster head, e.g., nodes with low energy, increases its chance to stay alive.

For that, we propose a sensor clustering method for dynamically organizing heterogeneous WSN using GA. Our method provides a framework to integrate multiple heterogeneity and clustering factors, which employs remaining energy, expected energy expenditure, network locality, and distance to the base station in search for an optimal, dynamic network structure for heterogeneous WSN. Heterogeneity factors are integrated as constraints to chromosomes and validation is performed to ensure network integrity. To avoid high energy consumption of sensor nodes, the base station will run the GA after each round to dynamically forming the structure of the network based on the new characteristics of the sensors, i.e., remaining energy. Section 2.2 presents the related work of constructing heterogeneous WSN to extend its lifetime. Then, Sect. 2.3 describes the proposed method for heterogeneous WSNs construction. Section 2.4 discusses our experimental results including a comparison study with five state-of-the-art methods and analysis of energy consumption. The chapter summary is finally described at Sect. 2.5.

2.2 Related Work

To extend the network lifetime in a heterogeneous network, Stable Election Protocol (SEP) [14] used weighted probabilities to elect cluster heads depending on the remaining energy in sensor nodes. Also, Developed Distributed Energy-Efficient Clustering (DDEEC) [12] method improved upon SEP by categorizing sensor nodes based on their energy level. The nodes with higher energy were the "advanced nodes", and the cluster head was selected from these group of nodes. Threshold Sensitive Stable Election Protocol (TSEP) [13] extended SEP method by grouping sensor nodes into three energy levels, and the cluster heads were selected based on thresholds.

Similarly, Energy Efficient Heterogeneous Clustered scheme (EEHC) [43] and Efficient Three Level Energy algorithm (ETLE) [44] selected cluster heads based on the probability proportional to the residual energy. In Hybrid Energy Efficient Reactive protocol (HEER) [45], the cluster head selection is based on the ratio of the residual energy of nodes and the average energy of the network. Both of Energy efficient heterogeneous clustered scheme (EEHC) [43] and Efficient Three Level Energy algorithm (ETLE) [44] assume three levels of heterogeneity and nodes are randomly distributed and are stationary. In EEHC, the cluster heads were selected based on weighted election probabilities of each node according to the residual energy. While in ETLE, each node chose a random number between 0 and 1. If the value of this random number was less than a threshold value, i.e., T, the node will be selected to serve as a cluster head. In Hybrid Energy Efficient Reactive protocol (HEER) [45], the CH selection is based on the ratio of the residual energy of nodes and the average energy of the network. All of these methods were proposed for WSN with initial energy heterogeneity only.

In [63], the Degree of connectivity is the main factor in selecting a CH. The degree of connectivity of a node, i.e., the number of its neighbors, is also a criterion that seems interesting to study. Intuitively, the more neighbors a sensor has, the more it seems to be an appropriate candidate as a cluster head, since a sensor with a low degree of connectivity might have little information, from its neighborhood, to aggregate and to forward to the BS. In the initial phase, each sensor is involved in the neighborhood information exchanges (hello protocol), which allows it to determine its degree of connectivity and the location of BS. In EEUC [64], the distance between the node and the BS is the main parameter for selecting the CH. The EEUC resulted in a network that is partitioned into clusters of unequal size, and that the clusters closer to BS have smaller sizes than those farther from the BS.

Many intelligent algorithms which provide adaptive methods that present intelligent behavior in complex and dynamic environments like WSNs exist [65]. Various researchers [65–69] debated the routing protocols in Cluster-based WSN based on intelligent algorithms as reinforcement learning, ant colony optimization, fuzzy logic, genetic algorithm, and neural networks. Furthermore, a lot of clustering mechanisms have been proposed. For example, Local Negotiated Clustering Algorithm presents a novel clustering method, which uses the similarity of nodes readings as an important feature during the process of creating a cluster. ACE construct the WSN clusters in a fixed number of iterations using the node degree as the main factor. In GA-WCA, load balanced factor with a sum of the distance from all neighbor nodes to CHs represents the main factor in network construction. On the other hand, LA2D-GA depends only on the distance as the main parameter to calculate fitness function that is used to evaluate the chance of the node to be a CH. LA2D-GA represent the chromosome in a two-dimensional grid which elucidates valid statistics of a WSN [70].

In [71], a two-level fuzzy logic approach is used to Cluster Head (CH) election based on four parameters namely—the number of neighbor nodes, remaining energy, energy dispersion and distance from the base station. The authors supported their idea for number of neighbors by stating that the number of neighbor nodes has been considered to be one determining parameter because CH must be chosen from an area where sufficient neighbor nodes are available LELE [72] protocol selects CH by remaining energy and the distance between a node and its neighbors, and the node with maximum energy and suitable position is chosen as the CH. LELE is proposed to improve load balancing in LEACH protocol Leader Election with Load balancing Energy. So, when the network is operating, the probability of the nodes becoming leader decreases or increases depending on the difference of the energy level of one node and neighbors, the distance of the node from neighbors, as well as the number of neighbors, and the probability of the nodes' to become a leader.

In [63], the Degree of connectivity is the main factor in selecting a CH. The degree of connectivity of a node, i.e., the number of its neighbors, is also a criterion that seems interesting to study. Intuitively, the more neighbors a sensor has, the more it seems to be an appropriate candidate as a cluster head, since a sensor with a low degree of connectivity might have little information, from its neighborhood, to aggregate and to forward to the BS. In the initial phase, each sensor is involved in the neighborhood information exchanges (hello protocol), which allows it to determine its degree of connectivity and the location of BS. In EEUC [64], the distance between the node and the BS is the main parameter for selecting the CH. The EEUC resulted in a network that is partitioned into clusters of unequal size, and that the clusters closer to BS have smaller sizes than those farther from the BS.

Searching for an optimal balance among many factors is non-trivial. Genetic Algorithm (GA) has been applied in the routing protocol of WSN [57, 58, 61]. When GA is used, a key objective is to define an appropriate fitness function that encodes the network structure and its goodness. However, most of the GA-based work was developed for the homogeneous model, i.e., HCR [62], while the remaining was concerned with static heterogeneous WSN. There is no additional efforts exist for the mobile heterogeneous model. The Evolutionary Based Clustered Routing Protocol (ERP) [57] overcame the limitations of clustering-algorithm-based genetic algorithms by uniting the clustering aspects of cohesion and separation error and proposed a new fitness function based on these two aspects.

2.3 Heterogeneous WSN Clustering Using Genetic Algorithm

2.3.1 Energy Model and Clustering Factors

As we deal with two levels of heterogeneity, our model has two types of sensor: normal and advanced sensor nodes. The advanced sensor has additional initial energy and lower energy consumption for data processing, i.e., receiving and transmitting messages. Based on that, we adopt the first order radio model to describe sensor energy [8] as shown in Fig. 2.1. The consumed energy E of a normal sensor node s is the summation of energy used to acquire l bits of data ($E_s^A(l)$), receive l' bits of data ($E_s^R(l')$), process l'' bits of data ($E_s^P(l'')$), and transmit l'' bits of data over a distance d ($E_s^T(l'', d)$):

$$E_s = E_s^A(l) + E_s^R(l') + E_s^P(l'') + E_s^T(l'', d), \tag{2.1}$$

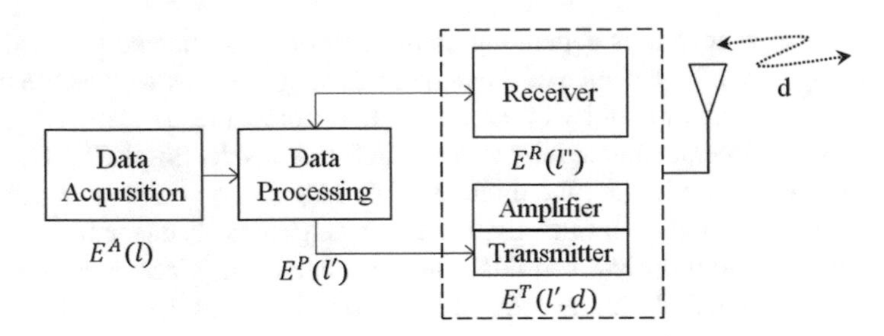

Fig. 2.1 First order radio model of a node. Each component has an energy consumption model that is a function of message length

where $E_s^R = E_i + l'E^*$ and E_i is the idle energy expenditure. $E_s^T = E_i + l''d^n$, and $n = 4$ for long distance transmission and $n = 2$ for short distance transmission, and E^* represents the cost of beam forming approach for energy reduction.

To compute the expected consumed energy $\hat{E}$ of a non-CH sensor node s' and a CH sensor node s, assume l bits of data are collected by each sensor node in a round. Given N_s sensors in a cluster, the expected consumed energy $\hat{E}$ are computed as follows:

$$\hat{E}_{s'} = E + lD^2(s', s), \tag{2.2}$$

$$\hat{E}_s = E + N_s l E^* + (N_s + 1) l D^4(s, B), \tag{2.3}$$

where E is the constant energy consumption including the energy of data acquisition, processing and idle. Functions $D(s', s)$ and $D(s, B)$ use Euclidean distance to give the distance between sensor nodes inside the cluster and from the cluster head to the base station, respectively.

The local sensor density is proportional to the number of sensors within the δ-vicinity as follows:

$$G_s(\delta) \propto \|S_s\|, \text{ and } S_s = \{s_i;\ D(s, s_i) \leq \delta\} \tag{2.4}$$

where S_s is the set of sensor nodes in the δ-vicinity of s and function $\|\cdot\|$ gives the set size.

2.3.2 Network Structure Building Using Genetic Algorithm

In our proposed GA-based framework, a binary chromosome is used to specify the CHs in the network, in which a one represents a CH and a zero represents a member node to a cluster as shown at Fig. 2.2. When a sensor becomes inactive, i.e., out of power, its corresponding gene value is set to -1, which exempts this sensor from further GA operations.

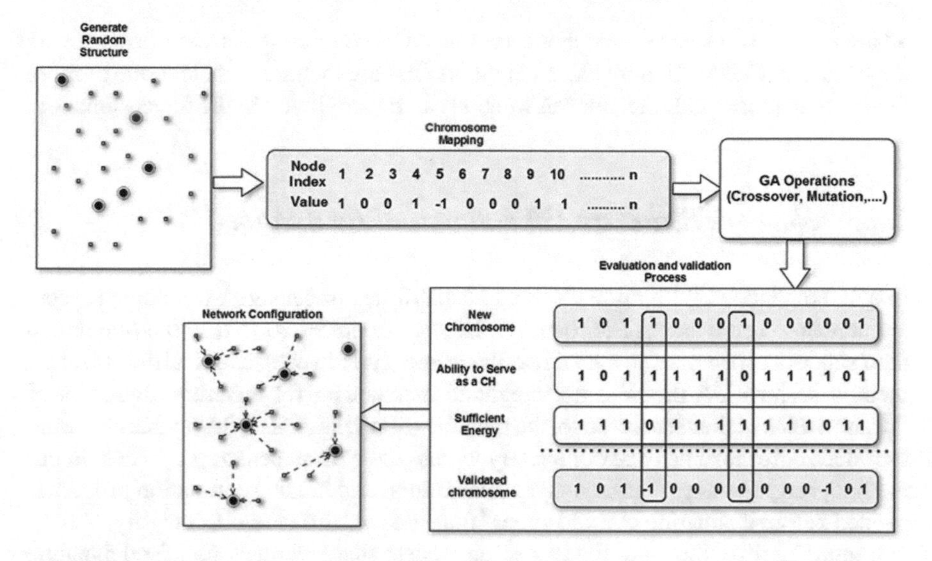

Fig. 2.2 Chromosome validation to ensure network integrity

The mapping to sensor clusters from a chromosome is to minimize the network communication distance $\mathbb{D}$ as follows:

$$\mathbb{D} = \sum_{i=1}^{C} \sum_{j=1}^{N_{s_i}} D(s_i, s_j) \tag{2.5}$$

where C is the number of clusters in a network and N_{s_i} is the number of member nodes in a cluster headed by node s_i. In practice, minimizing $\mathbb{D}$ is equivalent to assigning sensor nodes to clusters following the nearest neighbor rule.

The fitness function integrates energy factors, spatial distances, and the local sensor density:

$$f = \sum_{s} \frac{E_s(t)}{E_s(0)} + \frac{\tilde{E}}{\hat{E}} + \frac{1}{\hat{D}} + \frac{1}{N} \sum_{s'} G_{s'}(\delta), \tag{2.6}$$

where $E_s(t)$ is the remaining energy of sensor node s at round t and $E_s(0)$ is the initial energy of sensor node s. $\tilde{E}$ is the total energy cost if the messages are transmitted directly from all sensor nodes to the BS. $\hat{D}$ is the total distance between the CHs and the BS:

$$\hat{D} = \sum_{i=1}^{C} D(s_i, BS) \tag{2.7}$$

where each s_i is a sensor node that serves as a CH. Including sensor density favors the choice of CHs with more close neighbors. In cases where it is clear one or more factors play more vital role, uneven weights can be employed in the fitness function.

2.3.3 Network Structure Validation and Evaluation

In a heterogeneous WSN, functions, and capabilities of sensors vary. Some sensors are unable to serve as a cluster head, and some are preferred to take the role due to their superior processing power and available energy. However, classical optimization method such as GA provides no integrated mechanism for ensuring alignment of different roles of the sensors. Also, the random initialization and GA operations could introduce chromosomes that completely violate the current sensor properties. In our method, heterogeneity is presented as constraints, and hence a validation process is needed before evaluating chromosomes' fitness to ensure network integrity.

Figure 2.2 also illustrates the validation precess that leverages static and dynamic sensor properties. In the process of GA optimization, a new chromosome represents the proposed structure for the WSN. Each gene in the chromosome defines the expected role of the corresponding sensor node, i.e., whether it serves as a cluster head or a member node. The process consult 'the ability to serve as a CH', and 'the Sufficient Energy' tables. The role of 'The ability to serve as a CH' table is to determine whether the node can serve as a cluster head (one represents serving as cluster head; otherwise, member node). While, the 'Sufficient Energy' table is used to show the current energy status of the node, i.e., zero for disabled node and one for available node. The validation process determines if a chromosome complied with the constraints and hence retained in the offspring pool; otherwise, the chromosome is abandoned.

GA generates new chromosomes through crossover and mutation operations and evaluates their fitness. The crossover operation is performed with two randomly selected chromosomes determined by a crossover probability to regulate the operation. When crossover is determined not to be conducted, the parent chromosomes are duplicated to the offspring without change. Varying the crossover probability alters the evolution speed of the search process. In practice, the value of is close to 1.

The mutation operation involves altering the value at a randomly selected gene within the chromosome. Similarly, a mutation probability is used to regulate the performance of mutation. Different from the crossover probability, the mutation probability is usually fairly small. Essentially mutation operation could create completely new species, i.e., an arbitrary locus in the fitness landscape. Hence, it is a means to get out of a local optimum. Recall that when a sensor node becomes inactive, its corresponding gene is set to -1 to exempt it from mutation operations.

After validation process, Eq. (2.6) was used to evaluate the fitness of chromosomes. An intermediate pool of chromosomes was created to hold the individuals created in a generation, and depending on the needs user can specify any intermediate population size that is greater than the initial population size.

The evolution terminates when one of the following criteria is satisfied: (1) the maximum number of generations is reached; or (2) the fitness converges. Upon completion of the GA evolution, the chromosome that gives the best fitness value is used to restructuring the nodes.

2.3.4 Heterogeneous WSN Clustering Algorithm

Algorithm 1: Dynamic clustering of Heterogeneous WSN using Genetic Algorithm

1: Generate a pool of P chromosomes $U = \{u_1, u_2, \ldots, u_P\}$.
2: $\forall u_i \in U$, structuring the WSN by minimizing Eq. (2.5)
3: Evaluate the fitness of each $u_i \in U$ following Eqs. (2.6).
4: **for** $q = 1, 2, \ldots, Q$ **do**
5: $\quad \tilde{U} \Leftarrow \emptyset$
6: $\quad$ **for** $p = 1, 2, \ldots, P$ **do**
7: $\qquad$ Randomly select $u_a, u_b \in U$ $(a \neq b)$ based on the normalized fitness $\tilde{f}(u)$:

$$\tilde{f}(u) = \frac{f(u)}{\sum_c f(u)}.$$

8: $\qquad$ Cross over u_a and u_b according to α

$$\mathcal{C}(u_a, u_b|\alpha) \Rightarrow u'_a, u'_b.$$

9: $\qquad$ Perform mutation on u'_a and u'_b according to β

$$\mathcal{M}(u'_a|\beta) \Rightarrow \tilde{u}_a, \quad \mathcal{M}(u'_b|\beta) \Rightarrow \tilde{u}_b.$$

10: $\qquad$ Evaluate $f(\tilde{u}_a)$ and $f(\tilde{u}_b)$.
11: $\qquad$ $\tilde{U} \Leftarrow \tilde{U} \cup \{\tilde{u}_a, \tilde{u}_b\}$
12: $\quad$ **end for**
13: $\quad U \Leftarrow \{u_i; u_i \in \tilde{U} \text{ and } f(u_i)\}$
14: **end for**
15: Return the chromosome u^* that satisfies

$$u^* = \arg\max_u f(u), u \in U$$

Algorithm 1 presents our proposed method. In this algorithm, $q \in [1, Q]$ denotes the number of generations, and the population size is P. The pool of chromosome, denoted by U, is initialized with randomly generated individuals. An intermediate pool of chromosomes, denoted by $\tilde{U}$, is used to hold the individuals created in a generation, and depending on the needs user can specify any intermediate population size that is greater than the initial population size P. In crossover operation, two chromosomes are randomly selected from U and, according to the crossover probability α, two new chromosomes are created by switching consecutive genes. In mutation operations, the value of a randomly picked gene is altered between zero and one according to the mutation probability β.

2.4 Results and Discussion

In our evaluation, we assumed that each sensor node could directly reach the base station if it is provided with sufficient energy. The simulated sensor network was in an area of 100 m by 100 m (m) with 50 sensors randomly placed in the field, and the data packet size was 400 bits. The network parameters used in our experiments are listed in Table 2.1. The heterogeneity includes different initial power, data processing efficiency, and capability of serving as a cluster head. For the sensors with greater data processing efficiency, the energy used is 50% of that used by a regular sensor. 10% of sensor nodes possessed greater initial energy and data processing efficiency, and 10% of sensor nodes were unable to serve as a cluster head. The various sensors were chosen randomly in each experiment. Regarding GA running parameters, we used the population size of 30 for 30 generations. The crossover probability and mutation probability are 0.8 and 0.006, respectively. The neighborhood distance δ was 20 meters (m) throughout our experiments.

To evaluate SCHNGA in different environments, we created two cases with low and high sensors density, i.e., 50 node and 100 node, respectively. For each case, two scenarios of heterogeneity are designed: (1) sensors may differs in their initial energy, and (2) sensors may differs in initial energy, data processing capability, and

Table 2.1 Network parameters

Parameters		Values
Field area		100 m × 100 m
Base-station location		Center of the field
Energy of regular sensor		0.5 J
Energy of advanced sensor		1.0 J
Idle state energy		50 nJ/bit
Data aggregation energy		5 nJ/bit
Amplification energy	$d \geq d_0$	10 pJ/bit/m^2
(cluster head to base-station)	$d < d_0$	0.0013 pJ/bit/m^2
Amplification energy	$d \geq d_1$	$E_{fs}/10 = E_{fs1}$
(sensor to cluster head)	$d < d_1$	$E_{mp}/10 = E_{mp1}$
Packet size		400 bits
Percentage of advance sensors		0.1

Table 2.2 Network life span with different methods in the two proposed cases. FND: round at which first node die. LND: round at which last node die

Methods	100 sensor nodes		50 sensor nodes	
	FND	LND	FND	LND
ETLE	2040	8200	1514	6904
ERP	2756	10370	2010	9200
HEER	2340	7400	1789	6150
DDEEC	1879	10000	1100	8900
TSEP	2613	8000	1986	7640
SCHNGA	3510	12250	2690	10400
Improvement (SCHNGA) (%)	27.3	18.1	33.8	13

the ability to serve as a cluster head. Comparison studies were conducted with five state-of-the-art methods including HEER [45], TSEP [13], DDEEC [12], ETLE [44], and ERP [57].

Scenario 1: This scenario aims to evaluate the impact of heterogeneity regarding initial energy of the sensor nodes. Table 2.2 compares network life of our method with five state-of-the-art methods, which include HEER [45], TSEP [13], DDEEC [12], ETLE [44], and ERP [57]. The average number of rounds when the first node died (FND), and last node died (LND) is reported, and ten experiments were conducted for the analysis. Our method GAHN exhibited the longest average network life. The average improvement concerning the second best performance based on FND and LND is 33.8% and 13%, respectively. Figure 2.3 depicts the number of live nodes throughout the network life, which presents a progressive view. The dashed line with solid dot shows the results of GAHN. The balanced energy consumption greatly improved the network life and allowed the sensor energy to deplete evenly. This means the stability [57] of our method is the best one compared with the five other methods. It is clear that our proposed method greatly extended the network life.

Figure 2.3 depicts the change of the percentage of live sensor nodes throughout the entire network life. It is evident that the improvement of our proposed method is significant.

Scenario 2: The purpose of this scenario is to evaluate the impact of heterogeneity in terms of initial energy, processing capability, and the ability of the sensor to act as a cluster head. Table 2.3 presents the percentage of live sensor nodes throughout the life span of the WSN using the two cases for each scenario. The number of round is the average of 10 experiments with random sensor node placement. Compared with initial energy heterogeneity scenario, using the mentioned three factors of heterogeneity together yielded the largest number of rounds when the first sensor node dies. Depending on the sensor density, the improvement of using these three factors of heterogeneity was in the ranges of 20.8–38.4%. This means the more heterogeneity

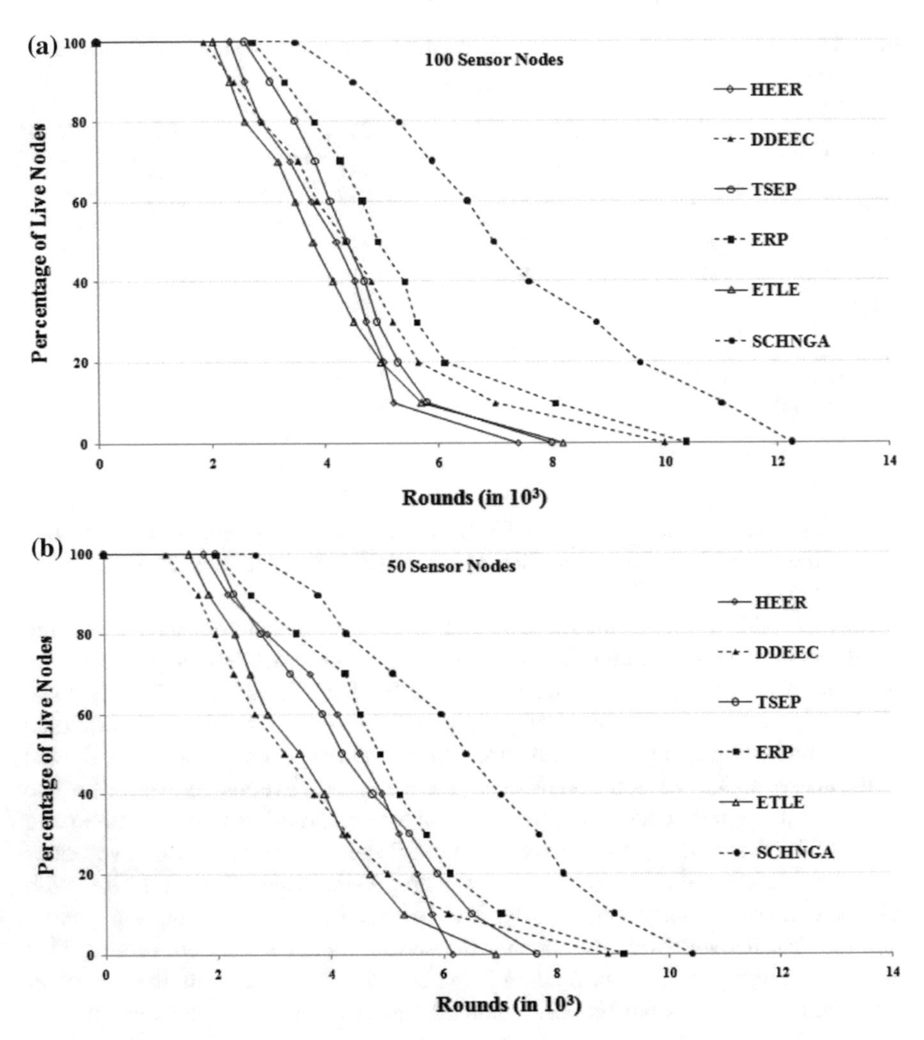

Fig. 2.3 Network lifetime for **a** high density field with 100 sensors and **b** low density filed with 50 sensors

capabilities assigned to the advanced nodes, the more network lifetime. It is clear that our proposed heterogeneity factors greatly extended the network lifespan.

Figure 2.8 depicts the change of the percentage of live sensor nodes throughout the entire network life. It is evident that the improvement of using different heterogeneity factors is significant. Our experiments also showed that the average number of clusters before the fist node die was 7% and 8% in high and low density fields respectively.

Figure 2.5 illustrates an example of the remaining energy at various transmission rounds of all sensors. At round 0, i.e., the initialization, 5 nodes (highlighted with green bars) were fueled with greater energy at 1 J. The red bars mark sensors unable

Table 2.3 Network life span with different heterogeneity factors in the two proposed cases. S-CH: some nodes can not serve as a clusters head

Heterogeneity factors	100 sensor nodes		50 sensor nodes	
	FND	LND	FND	LND
Energy	3510	12250	2690	10400
Energy, Processing, and S-CH	4860	19045	3250	14680

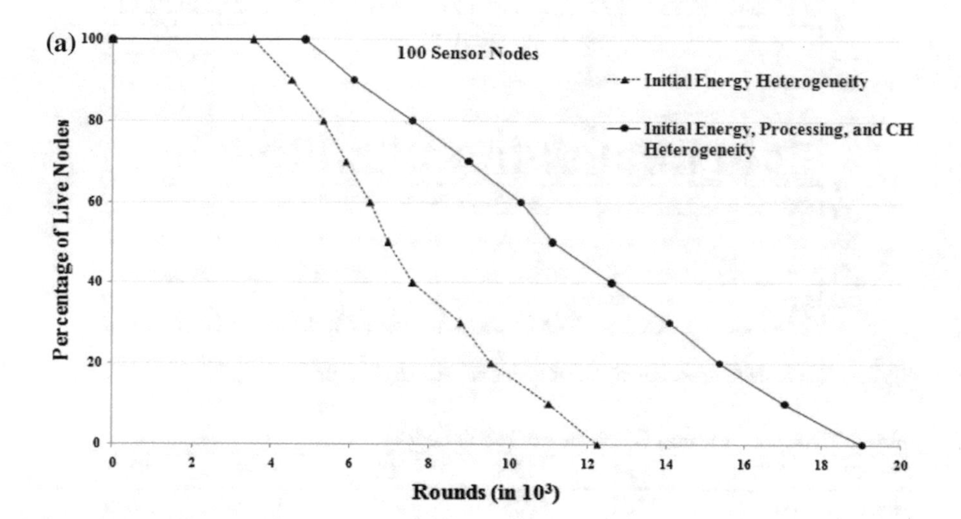

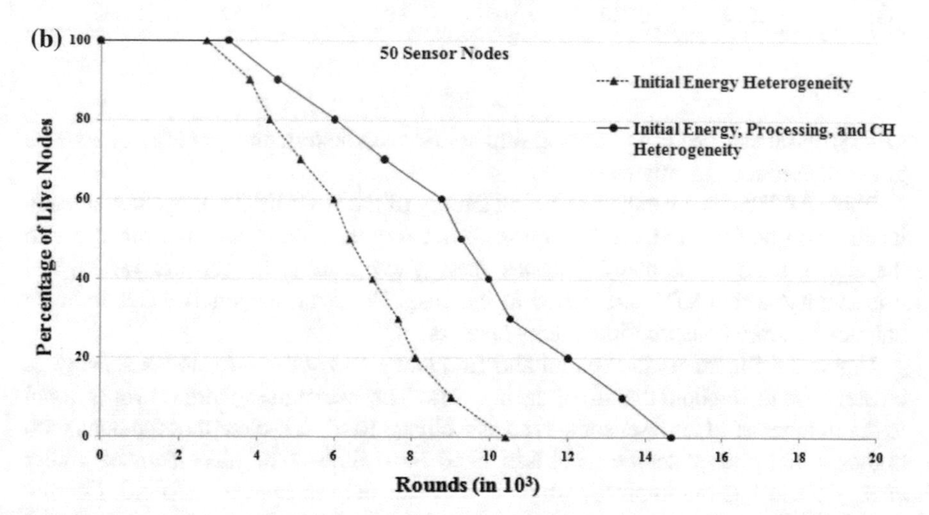

Fig. 2.4 Network lifetime for different terms of heterogeneity using **a** high density field with 100 sensors and **b** low density filed with 50 sensors

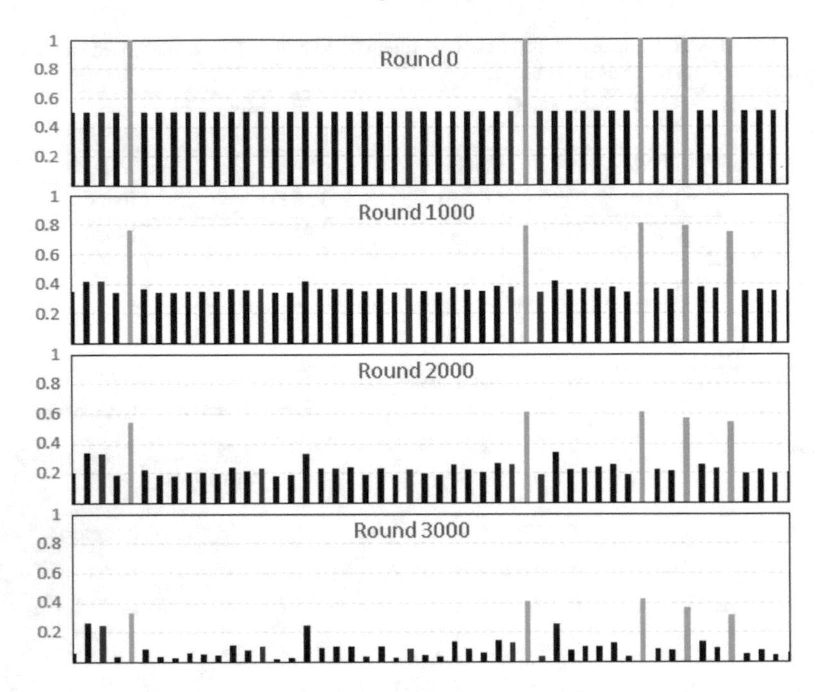

Fig. 2.5 The remaining energy of sensor nodes at various transmission rounds

Table 2.4 Remaining energy (J) variance of sensor nodes

Rounds	500	1000	1500	2000	2500	3000
Mean	0.431	0.363	0.295	0.226	0.158	0.090
STD	0.010	0.020	0.031	0.042	0.052	0.062

to serve as cluster head. As transmission continued, the remaining energy of sensors gradually reduced mostly evenly.

Table 2.4 lists the average remaining energy of the low-initial-energy sensors and its standard deviation at various transmission rounds. Due to unequal distances to the cluster head of the member nodes, energy expenditure for sensors varied, and it is inevitable that STDs continued to increase. However, the small STDs indicate balanced energy consumption among sensors.

Figure 2.6 illustrates the spatial and frequency view of sensor nodes serving as cluster head throughout the life of the network. The size of the sphere is proportional to the number of times a sensor served as a cluster head. It is clear that the ones with higher initial energy served as cluster head most times. The placement of higher energy sensors is randomized, which is unfortunately uneven in the field. Despite that, the high-initial-energy sensors dominated the choice of cluster head, their spatial disadvantage, i.e., closely located with each other, made some low-initial-energy sensors to act as cluster head to serve nearby sensors. The average number of clusters

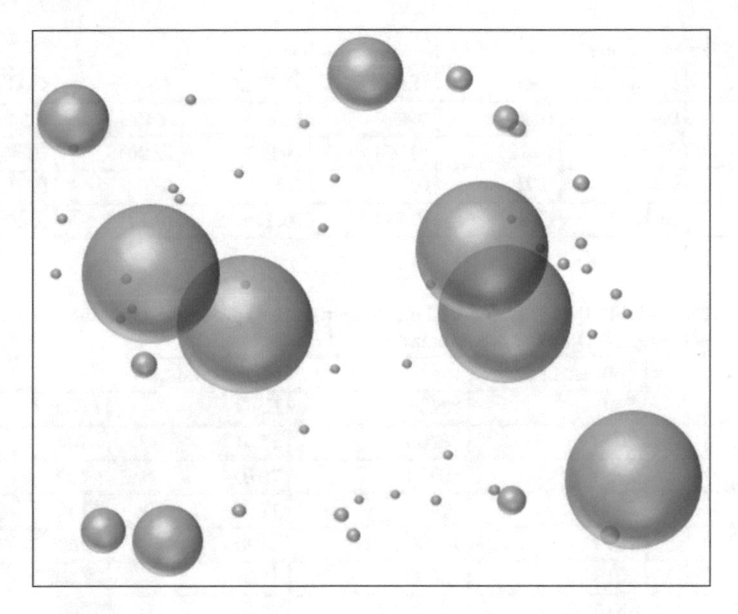

Fig. 2.6 Spatial and frequency view of sensor nodes serving as cluster head

in all rounds of our ten experiments is 6, among which 97% of times high-initial-energy nodes served as a cluster head. The forming of clusters was greatly influenced by the spatial location of sensor nodes. It is interesting to see that the low-initial-energy nodes that served as cluster head are usually far away from the high-initial-energy ones, which justifies their role as a cluster head.

Efficiency is an important factor in real-world applications [73]. Our experiments were conducted on a computer with Intel Core i5 2.6 GHz CPU, 4 GB memory, and Windows 7 operating system. The algorithms were implemented with a C# programming language. Table 2.5 lists the average time used to structure clusters in each transmission round. The time reported is before the first node became unavailable due to energy exhaustion. The number within parenthesis is the standard deviation. In addition to 50 sensors in the field, we also experimented with 100 randomly placed sensors with the other parameters remain the same. The average time used by GAHN was comparable to the other methods. **Note that the most time-consuming calculation in GAHN is evaluating the fitness of each chromosome, which can be implemented with parallel programming to improve efficiency in the future**.

Table 2.5 Average time (in seconds) for network structuring

Methods	ETLE	ERP	HEER	DDEEC	TSEP	GAHN
50	0.42	0.60	0.43	0.39	0.45	0.54
sensors	(0.03)	(0.12)	(0.02)	(0.04)	(0.06)	(0.06)
100	0.53	0.71	0.51	0.55	0.61	0.63
sensors	(0.10)	(0.34)	(0.21)	(0.11)	(0.17)	(0.27)

Table 2.6 Network life span with different methods in the two proposed scenarios. FND: round at which first node die. LND: round at which last node die

Methods	100 sensor nodes		50 sensor nodes	
	FND	LND	FND	LND
ETLE	2040	8200	1514	6904
ERP	2756	10370	2010	9200
HEER	2340	7400	1789	6150
DDEEC	1879	10000	1100	8900
TSEP	2613	8000	1986	7640
SCHNGA	3510	12250	2690	10400
SCHNGA-M	3105	10700	2208	9565
Improvement (SCHNGA) (%)	27.3	18.1	33.8	13
Improvement (SCHNGA-M) (%)	12.6	3.2	9.8	3.9

2.4.1 Another Experiments from Mobility Heterogeneity

To evaluate SCHNGA regarding mobility heterogeneity, we created two scenarios with low and high sensors density, i.e., 50 nodes and 100 nodes, respectively. For each scenario, two cases are designed: (1) the sensor nodes are stationary, and (2) the sensor nodes are mobile (in this case our proposed method is called 'SCHNGA-M'). Similarly to the previous experiments, the average performance of 10 repetitions is reported.

Table 2.6 presents the percentage of live sensor nodes throughout the lifespan of the WSN using different cases, i.e., static and mobile sensors. The number of the round is the average of 10 experiments with random sensor node placement. In the two cases of each scenario, our proposed method yielded the largest number of rounds when the first sensor node dies. Depending on the sensor density, the improvement was in the ranges of 27.3 to 33.44% using static sensors. While in case of sensors mobility, it was between 12.6 and 9.8%. This means the stability [57] of our method is the best one compared with the five other methods. It is clear that our proposed method greatly extended the network life.

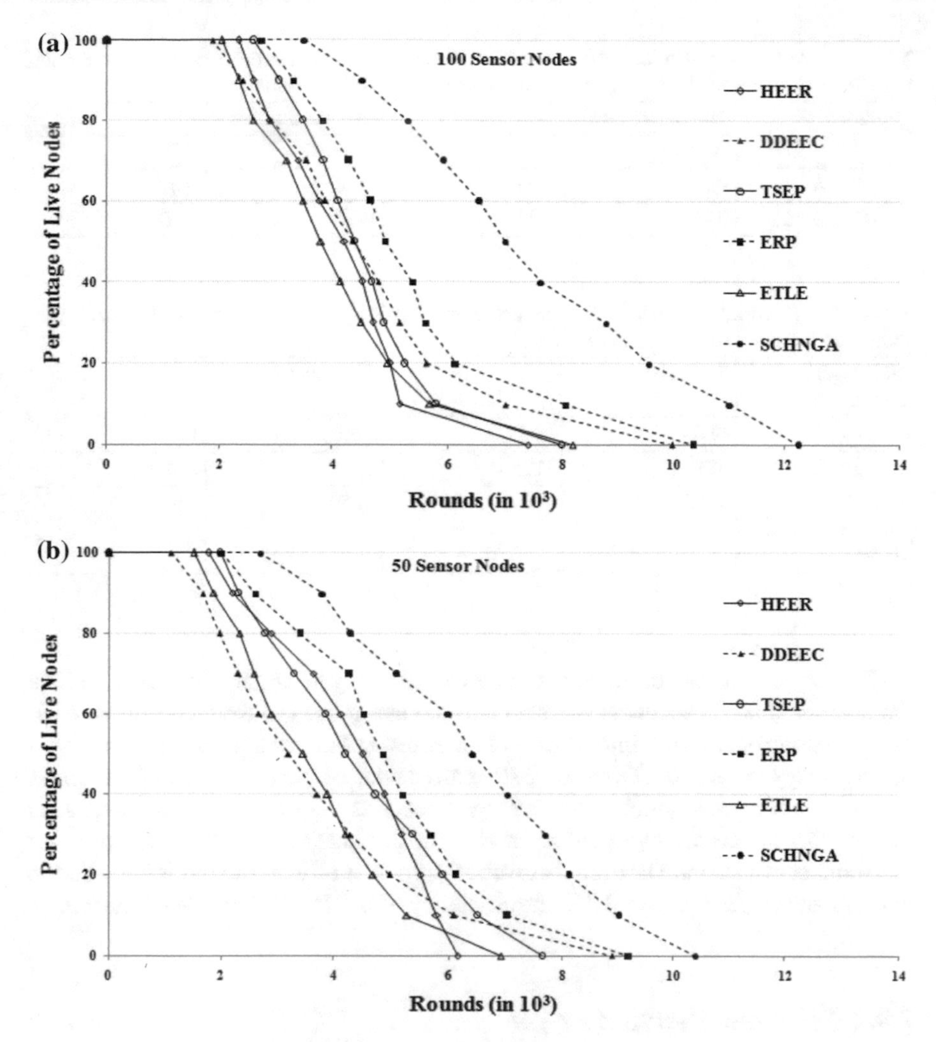

Fig. 2.7 Network lifetime for **a** high and **b** low density scenarios

Figure 2.7 depicts the change of the percentage of live sensor nodes throughout the entire network life. It is evident that the improvement of our proposed method is significant. Our experiments also showed that the average number of clusters before the fist node die was 8%, and 6% for a high-density scenario in case of static and mobile sensors respectively. While it was 9%, and 8% for a low-density scenario in case of static and mobile sensors respectively.

Table 2.7 Average time (in seconds) used to identify optimal network structure in each round using SCHNGA and SCHNGA-M compared with the five state-of-art methods

Methods	100 sensor nodes		50 sensor nodes	
	Average	STD	Average	STD
SCHNGA	0.63	0.27	0.54	0.06
SCHNGA-M	0.67	0.31	0.56	0.05

Table 2.8 Network life span with different heterogeneity factors in the two proposed cases. S-CH: some nodes can not serve as a clusters head

Heterogeneity factors	100 sensor nodes		50 sensor nodes	
	FND	LND	FND	LND
Energy	3510	12250	2690	10400
Energy, processing, and S-CH	4860	19045	3250	14680

To evaluate our method efficiency in case of mobility heterogeneity, Table 2.7 lists the average time (in seconds) and standard deviation used to form clusters in each transmission round. The time reported is before the first node became unavailable due to energy exhaustion. Despite the standard deviation increased when the number of sensor nodes was doubled, the average time was very close for all cases. It is evident that the efficiency of our method is mostly independent of sensor mobility and number of sensors. The overall average time across all experiments is 0.6 seconds with a standard deviation of 0.06. The efficiency of SCHNGA-M is also satisfactory.

2.4.2 *Different Version Again*

Table 2.8 presents the percentage of live sensor nodes throughout the life span of the WSN using the two cases for each scenario. The number of round is the average of 10 experiments with random sensor node placement.

Figure 2.8 depicts the change of the percentage of live sensor nodes throughout the entire network life. It is evident that the improvement of using different heterogeneity factors is significant.

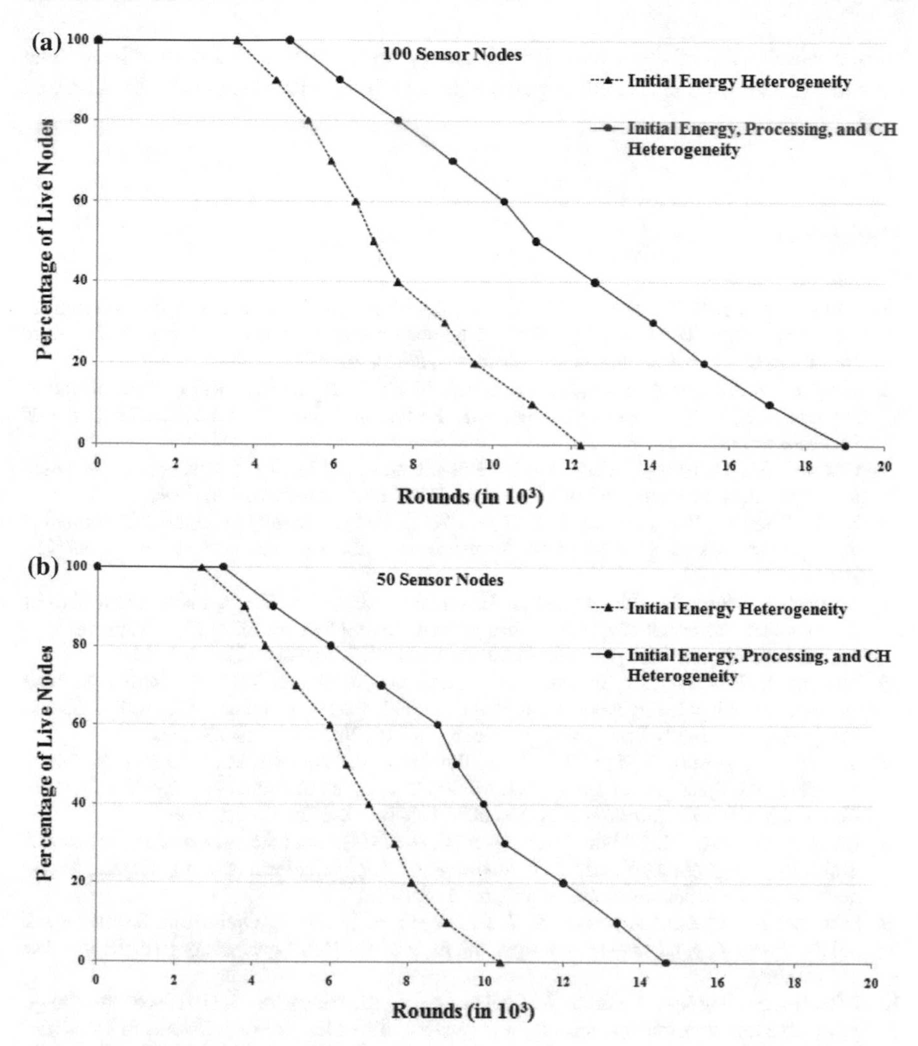

Fig. 2.8 Network lifetime for different terms of heterogeneity using **a** high density field with 100 sensors and **b** low density filed with 50 sensors

2.5 Conclusion

A self-clustering method for a heterogeneous network using Genetic Algorithm that optimizes the network clustering structure of sensor nodes is proposed. Compared with five state-of-the-art methods, the proposed method greatly extended the network life and the average improvement respect to the second best performance based on the first node, and the last node die was 33.8% and 13%, respectively. The results showed that the average number of clusters in all rounds of our experiments was 6, among which 97% of times high-initial-energy nodes served as a cluster head. The overall average time across all experiments was 0.6 s with a standard deviation of 0.06.

References

1. Elhoseny, M., Elminir, H., Riad, A., & Yuan, X. (2016a). A secure data routing schema for wsn using elliptic curve cryptography and homomorphic encryption. *Journal of King Saud University-Computer and Information Sciences*, *28*(3), 262–275.
2. Elhoseny, M., Farouk, A., Zhou, N., Wang, M., Abdalla, S., & Batle, J. (2017a). Dynamic multi-hop clustering in a wireless sensor network: Performance improvement. *Wireless Personal Communications*, 1–21.
3. Elhoseny, M., Tharwat, A., Farouk, A., & Hassanien, A. E. (2017b). K-coverage model based on genetic algorithm to extend WSN lifetime. *IEEE Sensors Letters*, *1*(4), 1–4.
4. Xie, D., Zhou, Q., You, X., Li, B., & Yuan, X. (2013). A novel energy-efficient cluster formation strategy: From the perspective of cluster members. *IEEE Communications Letters*, *17*(11), 2044–2047.
5. Tharwat, A., Mahdi, H., Elhoseny, M., & Hassanien, A. E. (2018). Recognizing human activity in mobile crowdsensing environment using optimized k-NN algorithm. *Expert Systems with Applications*. https://doi.org/10.1016/j.eswa.2018.04.017. Accessed 12 April 2018.
6. Tharwat, A., Elhoseny, M., Hassanien, A. E., Gabel, T., & Kumar, A. (2018). Intelligent Bezir curve-based path planning model using chaotic particle swarm optimization algorithm. *Cluster Computing*, (pp. 1–22). Springer. https://doi.org/10.1007/s10586-018-2360-3.
7. Sarvaghad-Moghaddam, M., Orouji, A. A., Ramezani, Z., Elhoseny, M., & Farouk, A. (2018). Modelling the Spice parameters of SOI MOSFET using a combinational algorithm. *Cluster Computing*. Springer. https://doi.org/10.1007/s10586-018-2289-6. (in Press).
8. Elhoseny, M., Yuan, X., El-Minir, H. K., & Riad, A. (2014). Extending self-organizing network availability using genetic algorithm. In *International conference on computing, communication and networking technologies (ICCCNT)*, pp. 1–6. IEEE.
9. Elhoseny, M., Tharwat, A., Yuan, X., & Hassanien, A. E. (2018). Optimizing K-coverage of mobile WSNs. *Expert Systems with Applications*, *92*, 142–153. https://doi.org/10.1016/j.eswa.2017.09.008.
10. Elhoseny, M., Farouk, A., Batle, J., Shehab, A., & Hassanien, A. E. (2017). Secure image processing and transmission schema in cluster-based wireless sensor network. In *Handbook of research on machine learning innovations and trends*, (Chapter 45, pp. 1022–1040), IGI Global. https://doi.org/10.4018/978-1-5225-2229-4.ch045.
11. Heinzelman, W. R., Chandrakasan, A., & Balakrishnan, H. (2000). Energy-efficient communication protocol for wireless microsensor networks. In *The Hawaii international conference on system sciences, Maui, Hawaii*.
12. Elbhiri, B., Rachid, S., & Elfkihi, S. (2010). Developed distributed energy-effecient clustering (DDEEC) for heterogeneous wireless sensor. In *Communications and mobile network*, (pp. 1–4). Rabat.
13. Kashaf, A., Javaid, N., Khan, Z., & Khan, I. (2012). TSEP: Threshold-sensitive stable election protocol for WSNs. In *Conference on Frontiers of information technology*, (pp. 164–168).
14. Smaragdakis, G., Matta, I., & Bestavros, A. (2004). SEP: a stable election protocol for clustered heterogeneous wireless sensor network. In *Second international workshop on sensor and actor network protocols and applications*.
15. Elhoseny, M., Abdelaziz, A., Salama, A. S., Riad, A. M., Muhammad, K., & Sangaiah, A. K. (2018). A hybrid model of internet of things and cloud computing to manage big data in health services applications. *Future Generation Computer Systems*. Elsevier. (in Press).

16. Abdelaziz, A., Elhoseny, M., Salama, A. S., & Riad, A. M. (2018). A machine learning model for improving healthcare services on cloud computing environment. *Measurement*, *119*, 117–128. https://doi.org/10.1016/j.measurement.2018.01.022.
17. Darwish, A., Hassanien, A. E., Elhoseny, M., Sangaiah, A. K., & Muhammad, K. (2017). The impact of the hybrid platform of internet of things and cloud computing on healthcare systems: Opportunities, challenges, and open problems. *Journal of Ambient Intelligence and Humanized Computing*. Springer. https://doi.org/10.1007/s12652-017-0659-1.
18. Yuan, X., Li, D., Mohapatra, D., & Elhoseny, M. (2017). Automatic removal of complex shadows from indoor videos using transfer learning and dynamic thresholding. *Computers and Electrical Engineering*. https://doi.org/10.1016/j.compeleceng.2017.12.026. (in Press).
19. Sajjad, M., Nasir, M., Muhammad, K., Khan, S., Jan, Z., Sangaiah, A. K., Elhoseny, M., & Baik, S. W. (2017). Raspberry Pi assisted face recognition framework for enhanced law-enforcement services in smart cities. *Future Generation Computer Systems*. Elsevier. https://doi.org/10.1016/j.future.2017.11.013.
20. Shehab, A., Elhoseny, M., El Aziz, M. A., & Hassanien, A. E. (2018). Efficient schemes for playout latency reduction in P2P-VoD systems. In A. Hassanien, & D. Oliva (Eds.), *Advances in soft computing and machine learning in image processing*. Studies in Computational Intelligence, Vol. 730. Cham: Springer. https://doi.org/10.1007/978-3-319-63754-9_22.
21. Elhoseny, M., Nabil, A., Hassanien, A. E., & Oliva, D. (2018). Hybrid rough neural network model for signature recognition. In A. Hassanien, & D. Oliva (Eds.), *Advances in soft computing and machine learning in image processing*. Studies in Computational Intelligence, Vol. 730. Cham: Springer. https://doi.org/10.1007/978-3-319-63754-9_14.
22. Abdeldaim, A. M., Sahlol, A. T., Elhoseny, M., & Hassanien, A. E. (2018). Computer-aided acute lymphoblastic leukemia diagnosis system based on image analysis. In A. Hassanien, & D. Oliva (Eds.), *Advances in soft computing and machine learning in image processing*. Studies in Computational Intelligence, Vol. 730. Cham: Springer. https://doi.org/10.1007/978-3-319-63754-9.
23. Elhoseny, H., Elhoseny, M., Riad, A. M., & Hassanien, A. E. (2018). A framework for big data analysis in smart cities. In A. Hassanien, M. Tolba, M. Elhoseny, & M. Mostafa (Eds.), *AMLTA 2018 the international conference on advanced machine learning technologies and applications (AMLTA2018)*. Advances in Intelligent Systems and Computing, Vol. 723. Cham: Springer. https://doi.org/10.1007/978-3-319-74690-6_40.
24. Elhoseny, M., Shehab, A., & Osman, L. (2018). An empirical analysis of user behavior for P2P IPTV workloads. In A. Hassanien, M. Tolba, M. Elhoseny, & M. Mostafa (Eds.), *AMLTA 2018 the international conference on advanced machine learning technologies and applications (AMLTA2018)*. Advances in Intelligent Systems and Computing, Vol. 723. Cham: Springer. https://doi.org/10.1007/978-3-319-74690-6_25.
25. Ehsan, S., Bradford, K., Brugger, M., Hamdaoui, B., Kovchegov, Y., Johnson, D., et al. (2012). Design and analysis of delay-tolerant sensor networks for monitoring and tracking free-roaming animals. *IEEE Transactions on Wireless Communications*, *11*(3), 1220–1227.
26. Elhoseny, M., Yuan, X., Yu, Z., Mao, C., El-Minir, H., & Riad, A. (2015). Balancing energy consumption in heterogeneous wireless sensor networks using genetic algorithm. *IEEE Communications Letters*, *19*(12), 2194–2197.
27. Iqbal, A., Akbar, M., Javaid, N., Bouk, S., Ilahi, M., & Khan, R. (2013). Advanced LEACH: A static clustering-based heterogeneous routing protocol for WSNs. *Journal of Basic and Applied Scientific Research*, *3*(5), 864–872.
28. Sudeep, T., Kumar, N., & Niu, J. (2014). EEMHR: energy-efficient multilevel heterogeneous routing protocol for wireless sensor networks. *International Journal of Communication Systems*, *27*(9), 1289–1318.
29. Elhoseny, M., Ramírez-González, G., Abu-Elnasr, O. M., Shawkat, S. A., Arunkumar, N., & Farouk, A. (2018). Secure medical data transmission model for IoT-based healthcare systems. *IEEE Access*, *PP*(99). https://doi.org/10.1109/ACCESS.2018.2817615.
30. Shehab, A., Elhoseny, M., Muhammad, K., Sangaiah, A. K., Yang, P., Huang, H., & Hou, G. (2018). Secure and robust fragile watermarking scheme for medical images. *IEEE Access*, *6*(1), 10269–10278. https://doi.org/10.1109/ACCESS.2018.2799240.

31. Farouk, A., Batle, J., Elhoseny, M., Naseri, M., Lone, M., Fedorov, A., Alkhambashi, M., Ahmed, S. H. & Abdel-Aty, M. (2018) Robust general N user authentication scheme in a centralized quantum communication network via generalized GHZ states. *Frontiers of Physics, 13*, 130306. Springer. https://doi.org/10.1007/s11467-017-0717-3.
32. Elhoseny, M., Elkhateb, A., Sahlol, A., & Hassanien, A. E. (2018). Multimodal biometric personal identification and verification. In A. Hassanien, & D. Oliva (Eds.), *Advances in soft computing and machine learning in image processing*. Studies in Computational Intelligence, Vol. 730. Cham: Springer. https://doi.org/10.1007/978-3-319-63754-9_12.
33. Elhoseny, M., Essa, E., Elkhateb, A., Hassanien, A. E., & Hamad, A. (2018). Cascade multimodal biometric system using fingerprint and Iris patterns. In A. Hassanien, K. Shaalan, T. Gaber, & M. Tolba (Eds.), *Proceedings of the international conference on advanced intelligent systems and informatics 2017, AISI 2017*. Advances in Intelligent Systems and Computing, Vol. 639. Cham: Springer. https://doi.org/10.1007/978-3-319-64861-3_55.
34. Wang, M. M., Qu, Z. G., & Elhoseny, M. (2017). Quantum secret sharing in noisy environment. In X. Sun, H. C. Chao, X. You, & E. Bertino (Eds.), *Cloud computing and security, ICCCS 2017*. Lecture Notes in Computer Science, Vol. 10603. Cham: Springer. https://doi.org/10.1007/978-3-319-68542-7_9.
35. Elsayed, W., Elhoseny, M., Riad, A. M., & Hassanien, A. E. (2018). Autonomic self-healing approach to eliminate hardware faults in wireless sensor networks. In A. Hassanien, K. Shaalan, T. Gaber, & M. Tolba (Eds.), *Proceedings of the international conference on advanced intelligent systems and informatics 2017, AISI 2017*. Advances in Intelligent Systems and Computing, Vol. 639. Cham: Springer. https://doi.org/10.1007/978-3-319-64861-3_14.
36. Abdelaziz, A., Elhoseny, M., Salama, A. S., Riad, A. M., & Hassanien, A. E. (2018). Intelligent algorithms for optimal selection of virtual machine in cloud environment, towards enhance healthcare services. In A. Hassanien, K. Shaalan, T. Gaber, & M. Tolba (Eds.), *Proceedings of the international conference on advanced intelligent systems and informatics 2017, AISI 2017*. Advances in Intelligent Systems and Computing, Vol. 639. Cham: Springer. https://doi.org/10.1007/978-3-319-64861-3_27.
37. Shehab, A., Ismail, A., Osman, L., Elhoseny, M., & El-Henawy, I. M. (2018). Quantified self using IoT wearable devices. In A. Hassanien, K. Shaalan, T. Gaber, & M. Tolba (Eds.), *Proceedings of the international conference on advanced intelligent systems and informatics 2017, AISI 2017*. Advances in Intelligent Systems and Computing, Vol. 639. Cham: Springer. https://doi.org/10.1007/978-3-319-64861-3_77.
38. Rizk-Allah, R. M., Hassanien, A. E., & Elhoseny, M. (2018). A multi-objective transportation model under neutrosophic environment. *Computers and Electrical Engineering*. Elsevier. https://doi.org/10.1016/j.compeleceng.2018.02.024. (in Press).
39. Batle, J., Naseri, M., Ghoranneviss, M., Farouk, A., Alkhambashi, M., & Elhoseny, M. (2017). Shareability of correlations in multiqubit states: Optimization of nonlocal monogamy inequalities. *Physical Review A, 95*(3), 032123. https://doi.org/10.1103/PhysRevA.95.032123.
40. El Aziz, M. A., Hemdan, A. M., Ewees, A. A., Elhoseny, M., Shehab, A., Hassanien, A. E., & Xiong, S. (2017). Prediction of biochar yield using adaptive neuro-fuzzy inference system with particle swarm optimization. In *IEEE PES PowerAfrica conference*, June 27–30, 2017. Accra-Ghana: IEEE. https://doi.org/10.1109/PowerAfrica.2017.7991209.
41. Ewees, A. A., El Aziz, M. A., & Elhoseny, M. (2017). Social-spider optimization algorithm for improving ANFIS to predict biochar yield. In *8th international conference on computing, communication and networking technologies (8ICCCNT)*, July 3–5, 2017. Delhi-India: IEEE.
42. Metawa, N., Elhoseny, M., Hassan, M. K., & Hassanien, A. E. (2016). Loan portfolio optimization using genetic algorithm: A case of credit constraints. In *Proceedings of 12th international computer engineering conference (ICENCO)*, (pp. 59–64). IEEE. https://doi.org/10.1109/ICENCO.2016.7856446.
43. Kumar, D., Aseri, T., & Patel, R. (2009). EEHC: Energy efficient heterogeneous clustered scheme for wireless sensor network. *Computer Communications, 32*(4), 662–667.
44. Tuah, N., Ismail, M., & Jumari, K. (2011). Energy efficient algorithm for heterogeneous wireless sensor network. In *IEEE international conference on control system and computing and engineering*, (pp. 92–96). Penang.

45. Javaid, N., Mohammad, N., Latif, K., Qasim, U., Khan, A., & Khan, M. (2013). HEER: hybrid energy efficient reactive protocol for wireless sensor networks. In *Saudi international electronics and communications and photonics conference*, (pp. 1–4). Riyadh.
46. Elhoseny, M., Yuan, X., El-Minir, H. K., & Riad, A. M. (2016b). An energy efficient encryption method for secure dynamic WSN. *Security and Communication Networks*, *9*(13), 2024–2031.
47. Elsayed, W., Elhoseny, M., Riad, A., & Hassanien, A. E. (2017). Autonomic self-healing approach to eliminate hardware faults in wireless sensor networks. In *International conference on advanced intelligent systems and informatics*, (pp. 151–160). Springer.
48. Elsayed, W., Elhoseny, M., Sabbeh, S., & Riad, A. (2017). Self-maintenance model for wireless sensor networks. *Computers and Electrical Engineering*. https://doi.org/10.1016/j.compeleceng.2017.12.022. (in Press).
49. Elhoseny, M., Yuan, X., El-Minir, H. K., & Riad, A. M. (2016). An energy efficient encryption method for secure dynamic WSN. *Security and Communication Networks*, *9*(13), 2024–2031. https://doi.org/10.1002/sec.1459.
50. Tripathi, K., Singh, N., & Verma, K. (2012). Two-tiered wireless sensor networks–base station optimal positioning case study. *IET Wireless Sensor Systems*, *2*(4), 351–360.
51. Wang, L., Wang, C., & Liu, C. (2009). Optimal number of clusters in dense wireless sensor networks: A cross-layer approach. *IEEE Transactions on Vehicular Technology*, *58*(2), 966–976.
52. Elhoseny, M., Tharwat, A., & Hassanien, A. E. (2017c). Bezier curve based path planning in a dynamic field using modified genetic algorithm. *Journal of Computational Science*. https://doi.org/10.1016/j.jocs.2017.08.004.
53. Metawa, N., Hassan, M. K., & Elhoseny, M. (2017). Genetic algorithm based model for optimizing bank lending decisions. *Expert Systems with Applications*, *80*, 75–82. https://doi.org/10.1016/j.eswa.2017.03.021.
54. Elhoseny, M., Shehab, A., & Yuan, X. (2017). Optimizing robot path in dynamic environments using genetic algorithm and Bezier curve. *Journal of Intelligent & Fuzzy Systems*, *33*(4), 2305–2316. IOS-Press. https://doi.org/10.3233/JIFS-17348.
55. Hosseinabadi, A. A. R., Vahidi, J., Saemi, B., Sangaiah, A. K., & Elhoseny, M. (2018). Extended genetic algorithm for solving open-shop scheduling problem. *Soft Computing*. https://doi.org/10.1007/s00500-018-3177-y).
56. Ali, P., Mashhadi, H., & Javadi, S. (2013). An optimal energy-efficient clustering method in wireless sensor networks using multi-objective genetic algorithm. *International Journal of Communication Systems*, *26*(1), 114–126.
57. Attea, B. A., & Khalil, E. A. (2012). A new evolutionary based routing protocol for clustered heterogeneous wireless sensor networks. *Applied Soft Computing*, *12*(7), 1950–1957.
58. Bayrakl, S., & Erdogan, S. (2012). Genetic algorithm based energy efficient clusters in wireless sensor networks. *Procedia Computer Science*, *10*, 247–254.
59. Elhoseny, Mohamed, Elleithy, Khaled, Elminir, Hamdi, Yuan, Xiaohui, & Riad, Alaa. (2015). Dynamic clustering of heterogeneous wireless sensor networks using a genetic algorithm towards balancing energy exhaustion. *International Journal of Scientific & Engineering Research*, *6*(8), 1243–1252.
60. Yuan, X., Elhoseny, M., El-Minir, H., & Riad, A. (2017). A genetic algorithm-based, dynamic clustering method towards improved WSN longevity. *Journal of Network and Systems Management*, *25*(1), 21–46.
61. Wu, Y., & Liu, W. (2013). Routing protocol based on genetic algorithm for energy harvesting-wireless sensor networks. *IET Wireless Sensor Systems*, *3*(2), 112–118.
62. Hussain, S., Matin, A., & Islam, O. (2007). Genetic algorithm for energy efficient clusters in wireless sensor networks. In *The 4th international conference on information technology ITNG*, (pp. 147–154). IEEE.
63. Diallo, C., Marot, M., & Becker, M. (2010). Single node cluster reduction in WSN and energy efficiency during cluster formation. In *The 9th annual mediterranean ad hoc networking conference, France*.

64. Chengfa, L., Mao, Y., Guihai, C., & Lie, W. (2005). An energy-efficient unequal clustering mechanism for wireless sensor networks. In *IEEE international conference on mobile ad hoc and sensor systems*, Washington, DC.
65. Guo, W., & Zhang, W. (2014). A survey on intelligent routing protocols in wireless sensor networks. *Journal of Network and Computer Applications*, *38*, 185–201.
66. Ahmed, G., Khan, N., & Ramer, R. (2008). Cluster head selection using evolutionary computing in wireless sensor networks. In *Progress in electromagnetics research symposium*, (pp. 883–886).
67. Asim, M., & Mathur, V. (2013). Genetic algorithm based dynamic approach for routing protocols in mobile ad hoc networks. *Journal of Academia and Industrial Research*, *2*(7), 437–441.
68. Bhaskar, N., Subhabrata, B., & Soumen, P. (2010). Genetic algorithm based optimization of clustering in ad-hoc networks. *International Journal of Computer Science and Information Security*, *7*(1), 165–169.
69. Karimi, A., Abedini, S., Zarafshan, F., & Al-Haddad, S. (2013). Cluster head selection using fuzzy logic and chaotic based genetic algorithm in wireless sensor network. *Journal of Basic and Applied Scientific Research*, *3*(4), 694–703.
70. Rana, K., & Zaveri, M. (2013). Synthesized cluster head selection and routing for two tier wireless sensor network. *Journal of Computer Networks and Communications*, *13*(3).
71. Kannammal, K., Purusothaman, T., & Manjusha, M. (2014). An efficient cluster based routing in wireless sensor networks. *Journal of Theoretical and Applied Information Technology*, *59*(3).
72. Shirmohammadi, M., Faez, K., & Chhardoli, M. (2009). LELE: leader election with load balancing energy. In *International conference on communications and mobile computing*, (pp. 106–110).
73. Elhoseny, M., Hosny, A., Hassanien, A. E., Muhammad, K., & Sangaiah, A. K. (2017). Secure automated forensic investigation for sustainable critical infrastructures compliant with green computing requirements. *IEEE Transactions on Sustainable Computing*, *PP*(99). https://doi.org/10.1109/TSUSC.2017.2782737.

Chapter 3
Hierarchical and Clustering WSN Models: Their Requirements for Complex Applications

Abstract Generally, WSN consists of thousands of inexpensive devices, called sensor nodes, capable of computation, communication and sensing events in a specific environment [1–3]. WSNs have attracted intensive interest from both academia and industry due to their wide application in civil and military scenarios [4–6]. Enormous advances that are emerging in WSNs act as a revolution in all aspects of our life. WSNs have unique specifications describe it and different from other networks. Sensor nodes have energy and computational challenges. Moreover, WSNs may be prone to software failure, unreliable wireless connections, malicious attacks, and hardware faults; that make the network performance may degrade significantly over time. Recently, there is a great interest related to routing process in WSNs using intelligent and machine learning algorithms such as Genetic Algorithms [7–9]. Security aspects in routing protocols have not been given enough attention, since most of the routing protocols in WSNs have not been designed with security requirements in mind [10–14]. In this chapter, the main models of WSN with their advantages and limitations are discussed, specially the clustering model. In addition, it provides a literature of the existing clustering methods of WSN that aims to increase the network lifetime. After that, the security aspects are explained in details. Finally, the existing secure clustering methods are discussed and evaluated based on a set of criteria.

3.1 WSN Concepts and Terminologies

3.1.1 Network Components

Recently, modern environments represent the next evolutionary development step in various utilities. Those environments need information about its surroundings as well as about its internal workings, so it resorted first to deal with sensory data from the real world for facilitation its tasks. Sensory data comes from multiple sensors of different modalities in distributed locations [21]. Smart technological advances, wireless sensor networks (WSNs) have attracted much interest in recent years [22–29]. They are now being considered for many critical applications [30–35] such as infrastructure monitoring, firefighting, pollution control, assisted living, military surveillance, for-

M. Elhoseny and A. E. Hassanien, *Dynamic Wireless Sensor Networks*, Studies in Systems, Decision and Control 165, https://doi.org/10.1007/978-3-319-92807-4_3

est monitoring, disaster management, space exploration, factory automation, secure installation, border protection, battlefield surveillance, tracking, industrial, building, shipboard, and transportation systems automation. Moreover, the dramatic reduction in the production cost of this wireless sensor has made its widespread deployment feasible. Hence, the urgent need for research into all aspects of WSNs has become evident. Now days, WSN has great, long-term potential for transforming our daily lives. It is evident to know that wireless sensor network is a self-organization network of small sensor nodes communicating among themselves using radio signals. These sensors are deployed in quantity to sense, monitor and understand the physical world. A WSN contains hundreds or thousands of sensor nodes, which are called motes. These sensors have the ability to communicate either among each other or directly to an external base-station (BS) [36]. A greater number of sensors allow sensing over larger geographical regions with a great accuracy. The sensors that are usually battery powered, deployed either randomly or according to a predefined statistical distribution over an environment. Although the individual sensors sensing range is limited, WSNs can cover a large space by integrating data from many sensors. Diverse and precise information on the environment may thus be obtained. Sensor networks are an emerging computing platform consisting of large numbers of small, low-powered, wireless motes. Each of them has limited computation, sensing and communication abilities. It is still a challenge to realize a distributed WSN comprising: small modules; high speed, low latency and reliable network infrastructures; software platforms are supporting easy and efficient installation of the WSN; and sensor information processing technologies. Sensor nodes are possibly prone to unexpected failures and malicious attacks. The main components of a general WSN are the sensor nodes, the sink (base station) and the events being monitored.

3.1.1.1 Sensor Node

In WSN, every sensor node has capabilities of sensing, processing and communicating data to the required destination. The basic entities in sensor nodes are sensing unit, power unit, processing unit and communication unit and memory unit to perform these operations. The role of the sensing unit is to create a connection between physical world and computation world. While memory unit is used to store both the data and program code. In order to store data packets from neighboring (other) nodes Random Only Memory (ROM) is normally used. And to store the program code, flash memory or Electrically Erasable Programmable Read Only Memory (EEPRM) is used. For computation and data transmission, the corresponding units in sensor node need power (energy). A node consist a power unit responsible to deliver power to all its units. The basic power consumption at node is due to computation and transmission where transmission is the most expensive activity at sensor node in terms of power consumption.

Figure 3.1 shows the design of main components of sensor nodes in wireless sensor networks which are based on the monitored environment, where the environment plays a key role in determining the size of the network, the deployment scheme and

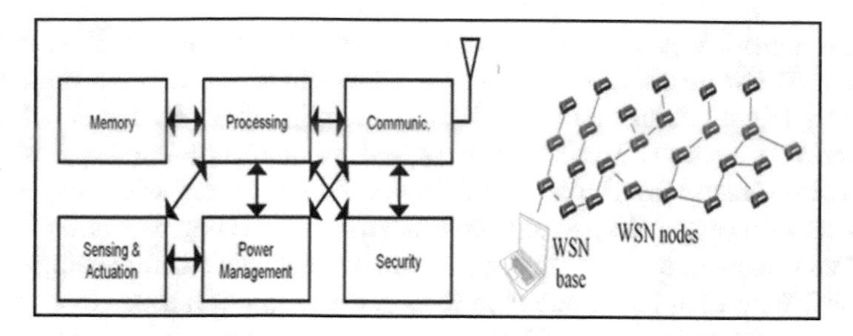

Fig. 3.1 The main components if a sensor node

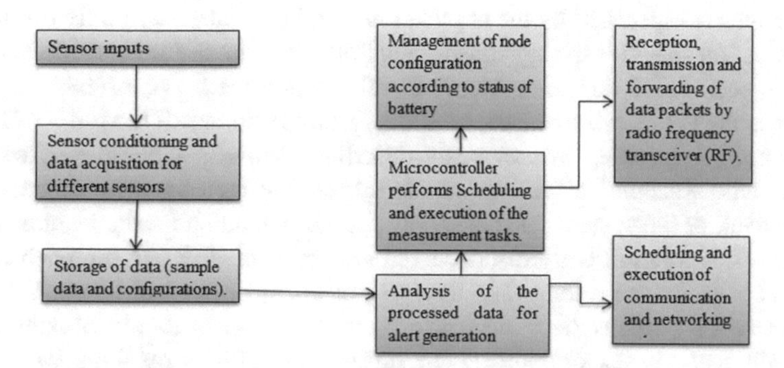

Fig. 3.2 Wireless sensor node functional block diagram

the network topology. The size of the network varies with the monitored environment. The basic issue in communication networks is the transmission of messages to achieve a prescribed message throughput and Quality of Service (QoS). QoS can be specified in terms of message delay, message due dates, bit error rates, packet loss, economic cost of transmission, transmission power, etc. QoS is Depending on the installation environment, economic considerations, and type of the topology.

Sensor, actuation and computation nodes are the fundamental components of distributed systems with wireless sensor nodes. To enable WSN-based applications, nodes (in general) have to provide the following basic functionality, Fig. 3.2:

- Signal conditioning and data acquisition for different sensors.
- Storage of data (sample data and configurations).
- Processing capabilities.
- Analysis of the processed data for alert generation.
- Actuation.
- Scheduling and execution of the measurement tasks.
- Management of node configuration (e.g. changing the sampling rate and reprogramming of data processing algorithms).
- Reception, transmission and forwarding of data packets.
- Scheduling and execution of communication and networking tasks.

Sensor nodes are possibly prone to unexpected failures and malicious attacks. Failures in WSN are volatile and may encounter crash or omission, timing, value or arbitrary failures. Crash failure is determined by a service that does not respond periodically to requests. This could be caused by radio interference that leads to occasional message loss. A crash or omission failure occurs when the service at some point stops responding to any request. An omission degree can be defined as the limited amount of omission failure that a node might have before being classified as crashed. While Timing failure represents services might fail due to a timeout in processing a request or by providing data too early. Timing failures occur when a node responds to a request with the correct value, but the response is received out of the time interval specified by the application. Timing failures will only occur when the application specifies timing constraints. Further Value failure considers services that have been failed due to an incorrect value when the service sends a response has inaccurate value. For instance, arbitrary failure includes all the types of failures that cannot be classified in previously described categories. Moreover, when the transient failure occurs, this means that the system reverts to its former status after a time interval, or the system fluctuates between normal and abnormal status. Hence, the network is intermittently disconnected and connected. However, each node, it remains connected for a long time interval uniformly at random range. When it is disconnected, it remains disconnected for a short time interval uniformly at random range. Although the sensor node may be suffered from faulty readings, which cause system's degradation, as opposite data readings consider necessary and related to the deployment time of sensor node. Therefore, the sensor that issues faulty readings are called faulty sensors. These faults can be attributed to: (i) faulty data measurement or data collection; (ii) some variables in the area surrounding the sensor have changed significantly; or (iii) the inherent performance of the sensor is abnormal. In sensor networks, an aggregation service could start sending both incorrect and correct values to the sink, or a node routing messages could not forward a message despite sending an acknowledge back to the sender. Such situations could be caused by malfunctioning software, hardware, corrupt messages, or even malicious nodes generating incorrect data.

3.1.1.2 The Basestation

The basestation is called the sink point to which all network data are directed. It is normally a resourceful node having unconstrained computational capabilities and energy supply. There can be single or multiple base stations in a network. Practically, the use of multiple base stations decreases network delay and performs better using robust data gathering. Base station in a network can also be stationary or dynamic. The dynamic base stations can influence the routing protocols greatly because of its changing position which will be not clear to all the nodes in a network. Beside mobility of base stations there are other characteristics of base stations like coverage, presence and number of nodes pose routing challenges for routing protocols which are explained in section.

3.1.1.3 Communication Unit

Senor nodes use radio frequencies or optical communication in order to achieve networking. This task is managed by radio units in sensor nodes that use electromagnetic spectrum to convey the information to their destinations. Usually each sensor node transfers the data to other node or sinks directly or via multihop routing.

3.1.2 Transmission Round

Transmission round can be defined as a specific time period through which the sensor nodes transmit data to the base station. It is used to determine the network life time and synchronize the data transmission to avoid data redundancy. The transmission round tasks depend on the network structure, i.e., static or dynamic. In the static network structure, the round starts by request from the basestation to sensor nodes to transmit data. Based on that, the sensor prepare the required data for transmission and choose the its destination, i.e., transmit the data directly to the basestation or to another node. The ability of the node to send the data depends on its remaining energy. If it has the sufficient energy, it sends the data to the destination and waits to get the acknowledgment. An example of the main operation at basestation and a sensor node during the transmission round is shown at Fig. 3.3 This figure assumes that the node has enough energy to transmit the data and it is connected directly to the basestation without intermediate gateway, i.e., cluster head.

3.1.3 Network Lifetime

There are more than one definition of the network lifetime. However, the main goal is to maximize it as long as possible. Mansy algorithms are proposed for this purpose as an optimization algorithms [37, 38, 40–44]. The most frequently used is the time at which the first node exhaust all its energy and become not available. This can be determined by the count of transmission rounds before the first node die. On the other hand, the network lifetime can be determined by a limited count of round based on the application in which the WSN is used. For example, in monitoring applications, the work of the network stop as soon as an event occurs and the sensors received it. While a few of researches assume that the network life time is determined by the round in which the last node of the network become unavailable.

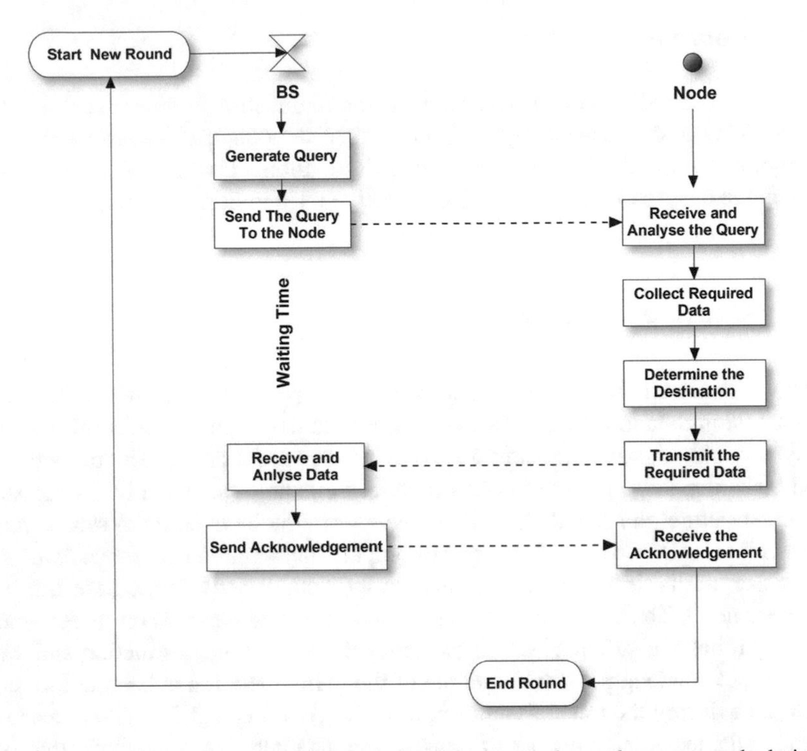

Fig. 3.3 The sequence of the operations at both of the basestation and a sensor node during the transmission round

3.1.4 The Difference Between WSN and the Traditional Networks

WSNs are usually deployed on land such as terrestrial WSNs that typically consist of hundreds to thousands of inexpensive wireless sensor nodes deployed in a given area. Those sensor nodes must be able to communicate effectively data back to the base station. As well, Multi-media WSNs that consist of a number of low cost sensor nodes equipped with cameras and microphones. These sensor nodes interconnect with each other over a wireless connection for data retrieval, process, correlation, and compression, for monitoring and tracking of events in the form of multimedia such as video, audio and imaging. Multi-media sensor nodes are deployed in a pre-planned manner into the environment to guarantee coverage. Also, Mobile WSNs consist of a collection of sensor nodes that can move on their own and interact with the physical environment, then spread out to gather information. The information gathered by a mobile node can be communicated to another mobile node when they are within the range of each other. Mobile nodes have the ability sense, compute, and communicate.

Moreover, WSNs have ability of deployment on underground and underwater environments. Although, the underground network consists of a number of sensor nodes buried underground or in a cave or mine and used to monitor underground conditions. Those nodes are expensive because appropriate equipment parts must be selected to ensure reliable communication through soil, rocks, water, and other mineral contents. The sink nodes in such networks are located above ground to relay information from the sensor nodes to the base station. An underground WSN is more expensive than a terrestrial WSN in terms of equipment, deployment, and maintenance. As opposite, underwater WSNs consist of a number of sensor nodes and vehicles deployed underwater. Those nodes are more expensive and fewer sensor nodes are deployed compared to underground WSNs.

Dependently, there are a number of differences between WSN and the traditional networks, i.e., Ad Hoc network; in applications, hardware and software limitations, and the mobility support. In fact, WSN is characterized by its strong interaction with its environment. For this reason, it can be used in a large number of applications such as environment monitoring, transportation applications. At the other side, the constraints in WSNs are stronger than in ad hoc networks. These constrains include both hardware and software limitations such as processing power, memory size, and energy consumption. The energy constraint represent the big challenge in WSN while it is not a affect the performance of the traditional networks. This is because, sensor are usually deployed in hostile environment. In addition, it is possible for sensor nodes to change their locations which is known as Node Mobility. Allowing the sensors to be mobile increases the number of WSN applications compared with stable sensors, i.e., tracking animal movements applications [45]. On the other hand, preventing some nodes to serve as a cluster head, e.g., nodes with low energy, increases its chance to stay alive. In traditional networks, the node location is specified in advance.

3.2 Wireless Sensor Network Challenges

WSNs are quickly gaining popularity due to the fact that they are potentially low cost solutions to a variety of real-world challenges. But there are many challenges for building an application using WSN. The common and traditional challenges include low power sensing, data acquisition, wireless transmission, energy source, and storage space. In addition, a set of challenges related to network operation and environment are exist. Figure 3.4 summarize these challenges. Traditionally, the majority of sensor based systems have been closed systems. For example, cars and airplanes have had networked sensor systems that operate largely within that vehicle. However, these systems and other WSN systems are expanding rapidly. WSN will enable an even greater cooperation and 2-way control on a wide scale: cars talking to each other and controlling each other to avoid collisions, humans exchanging data automatically when they meet and this possibly affecting their next actions, and physiological data uploaded to doctors in real-time with real-time feedback from the doctor. WSN require openness to achieve these benefits. However, supporting openness creates

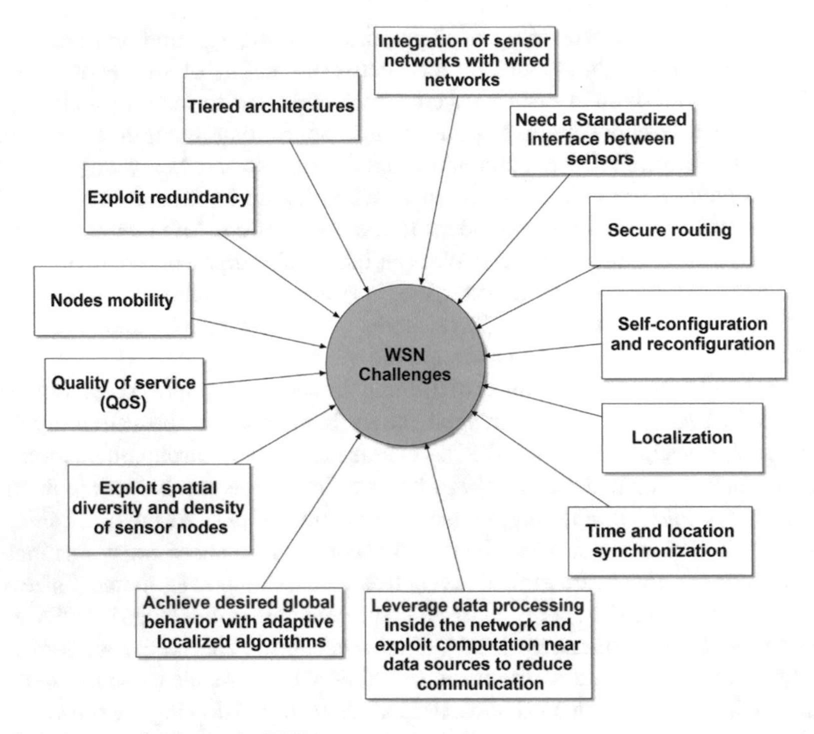

Fig. 3.4 Current challenges for WSN

many new research problems including dealing with heterogeneity. All of our current composition techniques, analysis techniques, and tools need to be re-thought and developed to account for this openness and heterogeneity [46]. New unified communication interfaces will be required to enable efficient information exchange across diverse systems and nodes. Of course, openness also causes difficulty with security and privacy, the topics of the next two subsections. Consequently, openness must provide a correct balance between access to functionality and security and privacy [47].

In addition, Self-organizing is an important feature for many applications in WSN after deployment specially in clustering model. At the conclusion of the self-organizing stage it is common for the nodes of the WSN to have synchronized clocks, know their neighbors, and have a coherent set of parameter settings such as consistent sleep/wake-up schedules, appropriate power levels for communication, and pair-wise security keys [46]. For example, some nodes may be physically moved unexpectedly. More and more nodes may become out of place over time. To make system-wide node locations coherent again, node re-localization needs to occur.

Moreover, a fundamental problem that must be solved in WSN is dealing with security attacks [48]. Security attacks are problematic for WSN because of the minimal capacity devices being used in parts of the systems, the physical accessibility to sensor and actuator devices, and the openness of the systems including the fact that most devices will communicate wirelessly. The security problem is further exacerbated because transient and permanent random failures are commonplace in WSN and failures are vulnerabilities that can be exploited by attackers. However, the considerable redundancy in WSN creates great potential for designing them to continue to provide their specified services even in the face of failures. To meet realistic system requirements that derive from long lived and unattended operation, WSN must be able to continue to operate satisfactorily in the presence of, and to recover effectively from security attacks. The system must also be able to adapt to new attacks unanticipated when the system was first deployed.

On the other hand, the ubiquity and interactions of WSN provide many conveniences and useful services for individuals, but also create many opportunities to violate privacy [49]. To solve the privacy problem created by single and interacting WSN of the future, the privacy policies for each (system) domain must be specified. Once specified the WSN system must enforce privacy. Consequently, the system must be able to express users requests for data access and the systems policies such that the requests can be evaluated against the policies in order to decide if they should be granted or denied. One of the more difficult privacy problems is that systems may interact with other systems, each having their own privacy policies. Consequently, inconsistencies may arise across systems. Once again, on-line consistency checking and notification and resolution schemes are required.

Further, WSN often support many real-time sensor streams in redundancy, uncertain, and open environments. In particular, a very difficult issue is that wireless communication packet delivery is subject to burst losses. New concepts of guarantees must be developed that will likely span a spectrum from deterministic to probabilistic depending on the application, the environment, and noise and interference models [47].

3.3 Routing Models

Different routing protocols are designed to fulfill the shortcomings of the recourse constraint nature of the WSNs. The deployed WSN can be differentiated according to the network structure or intended operations. Therefore, routing protocols for WSN needs to be categorized according to the nature of WSN operation and its network architecture. WSN routing protocols can be subdivided into two broad categories, network architecture based routing protocols, i.e., flat and clustering models; and operation based routing protocols, i.e., proactive and reactive models. Here, the network architecture based routing protocols are discussed as they are the popular models in most applications. In addition, they have a positive effect on the network energy and lifetime.

3.3.1 Flat Routing Model

The traditional routing architecture in WSN is called flat routing mode. In this model, each sensor node has a direct connection with the base-station as shown at Fig. 3.5. In addition, all sensors are typically assigned same roles in gathering information. At each transmission round, each node collects the required information and sends it directly to the base-station if it has sufficient energy. This model works well in case of static network structure in which the topology of the network will not be changed during the network lifetime. This is because there is no need to reconstructing the network structure in case of node die. On the other side, the network loses a lot of energy as a result of direct data transmission between node to the base-station every round. With different words, the flat routing architecture is called 'Address-centric routing (AC)'. In AC, each source independently sends data along the shortest path to sink based on the route that the queries took (end-to-end routing). An example is shown at Fig. 3.6 in which each source sends its information separately to the sink.

3.3.2 Clustering Routing Model

To enhance the network life time by the period of a particular mission, many routing protocols have been devised. One of these are network clustering, in which network is partitioned into small clusters and each cluster is monitored and controlled by a node, called Cluster Head (CH). The CH should be powerful, closer to the cluster-centroid,

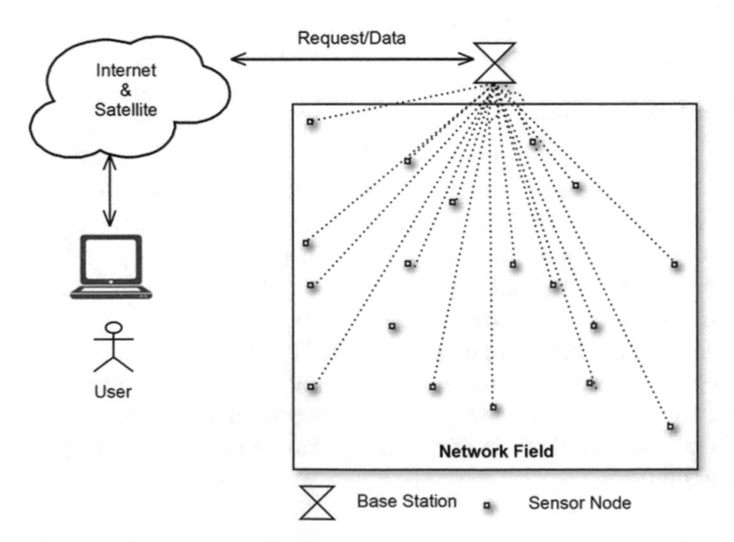

Fig. 3.5 The traditional flat routing architecture in which each node transmit data directly to the basestation

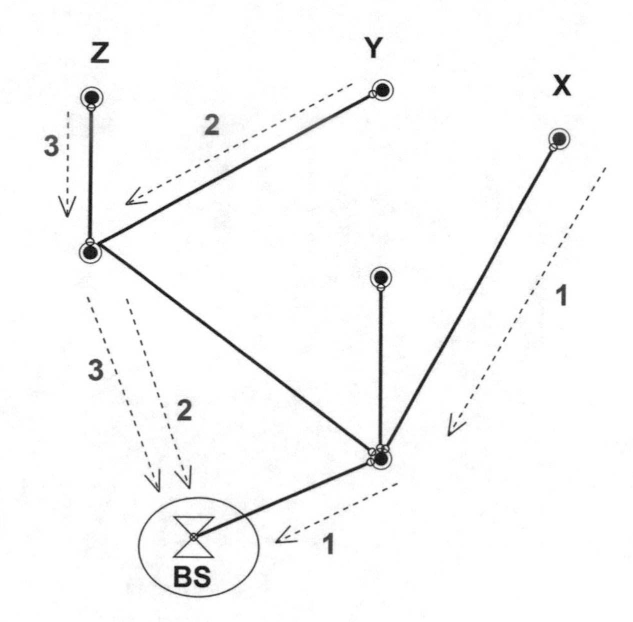

Fig. 3.6 The an example of address-centric routing model

less vulnerable and has low mobility, so that it can aggregate the data from its own cluster nodes and then send it directly to the Base Station (BS) [50]. WSN topologies are classified into four types of models (Single-hop flat model, Single-hop clustering model, Multi-hop flat model, and Multi-hop clustering model) [51]. As shown at Fig. 3.7, in the single-hop models, all sensor nodes transmit their data to the BS directly. These architectures are infeasible in large-scale areas because transmission cost becomes expensive in terms of energy consumption and in the worst case, the BS may be unreachable. In the multi-hop flat model, because all nodes should share the same information such as routing tables, overhead and energy consumption can be increased.

On the other hand, in the multi-hop clustering model, sensor nodes can maintain low overhead and energy consumption because particular cluster heads aggregate data and transmit them to the BS. Additionally, wireless medium is shared and managed by individual nodes in the multi-hop flat model, which results in low efficiency in the resource usage. In the multi-hop clustering model, resources can be allocated orthogonally to each cluster to reduce collisions between clusters and be reused cluster by cluster. As a result, the multi-hop clustering model is appropriate for the sensor network deployed in remote large-scale areas. In this work, the multi-hop cluster model will be selected.

There are various problems in cluster management process. A major challenge in WSNs is to select appropriate cluster heads in the cluster based model. Choosing the CH will affect the network lifetime because of collecting the data from its cluster nodes and forward it to base station require a special characteristics of the CH. Moreover, by rotating the cluster-head, energy consumption is expected to be uniformly distributed among all nodes.

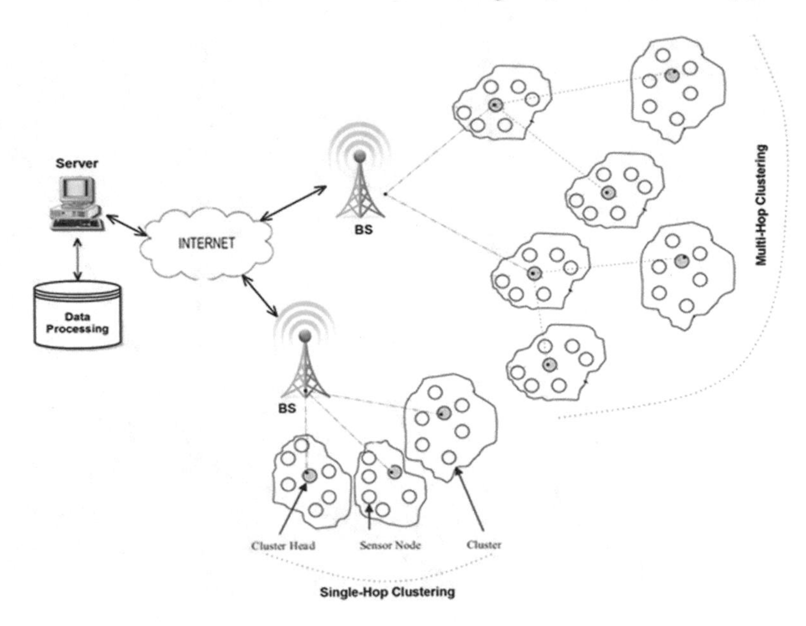

Fig. 3.7 WSN clustering model

The clustering model is called 'Data-centric Model (DC)'. In DC, the sources send data to the sink, but routing nodes enroute look at the content of the data and perform some form of aggregation/consolidation function on the data originating at multiple sources as shown at Fig. 3.8. The data from the sources is aggregated at the aggregation point, and the combined data (labelled 1 + 2 + 3) is sent from the sink. The latter results in energy savings as fewer transmissions are required to send the information from the sources to the BS.

3.4 WSN Application Areas

Wireless sensor networks (WSN) is an important and exciting new technology with great potential for improving many current applications in medicine, transportation, agriculture, industrial process control, and the military as well as creating new revolutionary systems in areas such as global-scale environmental monitoring, precision agriculture, home and assisted living medical care, smart buildings and cities, and numerous future military applications [47].

Now, they are used in many industrial and civilian areas. WSN has a wide range of applications in the fields such as military, industrial, domestic, and environmental areas. WSN are used in the military for communication, surveillance, reconnaissance, computing and target tracking. It can also be used for monitoring friendly and opposing forces. For example, it can be used to assess damage caused in a battle, to

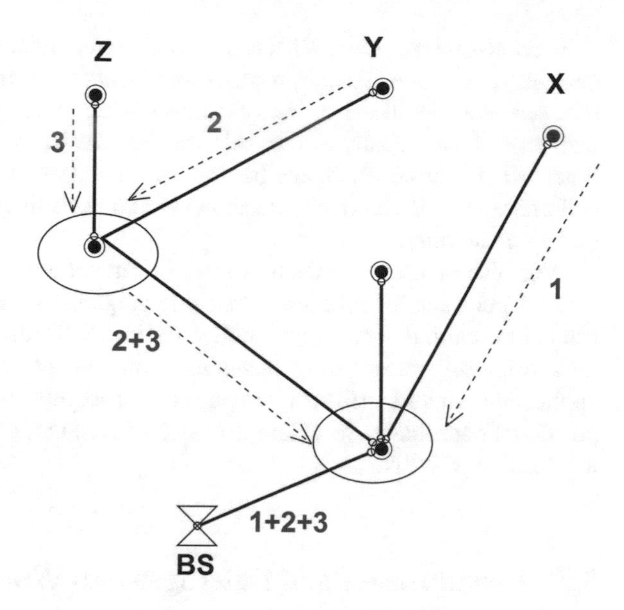

Fig. 3.8 The an example of data-centric routing model

detect chemical, biological or nuclear attack. They can be used for battlefield surveillance by sensing acoustic and magnetic signals produced by target objects such as vehicles and troop movements. WSN is a promising technology in the military due to its characteristics like rapid deployment, error diagnosis and self-organization. During war, they reduce casualties providing critical reaction time for the troops to react. WSN can also be used for safety purposes like reconnaissance of nuclear radiation without a soldier getting exposed to it. There are ongoing research studies to identify a shooters location using sensor networks. Here, the sensor nodes measure the muzzle blast and shock waves created by the moving bullet to identify the exact location of the shooter.

WSN are used in environmental applications for tracking animals, birds and environmental conditions that affect crops. It can be used to monitor soil conditions for irrigation, fertilization and harvesting. WSN are used to detect forest fires, floods and volcanic activity. To detect a forest fire, many sensor nodes are deployed in a forest at high density. These sensor nodes can detect the exact location of the origin of the fire. This helps fire fighters to prevent fire from spreading uncontrollably. WSN can be used for monitoring the air for pollution by attaching gas sensors to mobile vehicles or fixed locations on the street. WSN can also used to monitor permafrost to study the slope stability to avoid natural hazards.

In the health industry [52], sensor networks can be used to monitor doctors, patients and drug administrators. They can be used in hospitals to monitor physiological signals like blood pressure, blood glucose level, body temperature, pulse oxygen saturation, electrocardiogram, etc. They can also be used to monitor internal infection processes of a human body. Attaching sensor nodes to medications, could avoid administration of wrong medications to patients with specific allergies.

For domestic purposes, WSN can be used to create smart home environment where devices like vacuum cleaners, ovens, disk players and refrigerators can be connected with sensor nodes. These devices can thus interact with each other and communicate with an end user locally or remotely through the internet. One such application is Smart Kindergarten which can be used to solve developmental problems of young children where they learn by interacting with objects like toys attached with sensors, in their environment.

WSN can be used in industry to monitor material fatigue, product quality, wind tunnels, guidance for robots and machine diagnosis to name a few. WSNs are also used in the car industry for automotive testing. WSN can also be used for real time monitoring of nuclear power plants for radiation and faults [12, 13]. All of these applications to be feasible, the system might require the proper reporting of their positional coordinates to obtain a meaningful outcome. This can be achieved by localization of WSN.

3.5 Homogeneous and Heterogeneous WSN

In homogeneous network all the nodes are identical in terms of battery energy and hardware complexity. Homogeneous networks have pure static clustering. In homogeneous type of network, single network topology is used and this is not a complex network. Thus in short we can say that, sensor network in which the sensor nodes having the similar hardware complexity and battery energy is called as homogeneous sensor network. But the most important drawback of homogeneous sensor network as well as role rotation is that all network nodes will be able to act as cluster heads, and hence they should posses the required capabilities related to hardware requirements.

On the other side, heterogeneous network different topologies are used and this makes the network a very complex network. Thus in short, we can say that in case of heterogeneous sensor network there are two or more various types of network nodes along with different functionality and battery energy is used. The real motivation behind the heterogeneous networks is the need of extra battery energy and more complex hardware is embedded in some cluster heads, hence this reducing the overall cost of hardware for the remaining sensor network. But the fixing of cluster head nodes is nothing but the role rotation which is not possible longer.

In both homogeneous and heterogeneous WSN, there are number of design factors for designing an effective and efficient wireless sensor networks. Some of them are illustrated below:

1. Self-healing: WSNs are different from traditional networks. Nodes in WSNs are prone to failure due to energy depletion, hardware failure, communication link errors, malicious attack, and so on. If its operating quality decreases at all, a small failure can cause total breakdown. Therefore, the management system should be aware of such dynamics and it should adjust accordingly.

2. Memory and Processing Limitations: The sensor nodes are supposed to have limited memory and processing power. The management applications need to be aware of such constraints and may only impose minimal overhead on the low-powered nodes for the storage of management information and processing.
3. Limited Bandwidth Consumption: The energy cost associated with communication is usually more than that of sensing and processing. Therefore, the management applications should be designed with this consideration in mind. Moreover, some sensor technologies may also have bandwidth limitations in the presence of high channel impairments.
4. Adaptability: The management system should be able to adjust to network dynamics and rapid changes in the network topology. The system should be able to gather the reported state of the network and the topology changes. It should also be able to handle node mobility, the addition of new nodes, and the failure of existing nodes.
5. Scalability: Sensor nodes are assumed to be deployed in large numbers. The management system should be able to handle large volumes of sensory data as well as high density of sensor nodes.
6. Production Costs: The cost of each sensor node has to be kept low at least by allocating minimal resources to the management system.
7. Power Consumption: Sensor devices are mostly battery-operated; therefore, the management applications should be able to run on the sensor nodes without consuming too much energy. The management operations should be lightweight on node-local resources in order to prolong its lifetime, thereby, contributing to the network lifetime as a whole.

3.6 Conclusion

Wireless Sensor Networks (WSNs) are extensively used in different applications. However, there are may challenges that must be addressed to design a robust WSN-based model such as the network lifetime. This chapter provides a detailed review about WSN terminologies and design challenges. In both homogeneous and heterogeneous WSN, the design factors for designing an effective and efficient are described. A comparison between the traditional networks and WSN are explained too.

References

1. Elhoseny, M., Farouk, A., Zhou, N., Wang, M., Abdalla, S., & Batle, J. (2017a). Dynamic multi-hop clustering in a wireless sensor network: Performance improvement. *Wireless Personal Communications*, 1–21.
2. Elhoseny, M., Yuan, X., El-Minir, H. K., & Riad, A. (2014). Extending self-organizing network availability using genetic algorithm. In *International conference on computing, communication and networking technologies (ICCCNT)*, pp. 1–6. IEEE.

3. Elhoseny, M., Yuan, X., Yu, Z., Mao, C., El-Minir, H., & Riad, A. (2015). Balancing energy consumption in heterogeneous wireless sensor networks using genetic algorithm. *IEEE Communications Letters*, *19*(12), 2194–2197.
4. Elhoseny, M., Tharwat, A., Farouk, A., & Hassanien, A. E. (2017b). K-coverage model based on genetic algorithm to extend WSN lifetime. *IEEE Sensors Letters*, *1*(4), 1–4.
5. Elhoseny, M., Tharwat, A., Yuan, X., & Hassanien, A. E. (2018). Optimizing K-coverage of mobile WSNs. *Expert Systems with Applications*, *92*, 142–153. https://doi.org/10.1016/j.eswa.2017.09.008)
6. Elhoseny, M., Farouk, A., Batle, J., Shehab, A., & Hassanien, A. E. (2017). Secure image processing and transmission schema in cluster-based wireless sensor network. In *Handbook of research on machine learning innovations and trends*, (Chapter 45, pp. 1022–1040), IGI Global, 2017. https://doi.org/10.4018/978-1-5225-2229-4.ch045.
7. Elhoseny, M., Tharwat, A., & Hassanien, A. E. (2017c). Bezier curve based path planning in a dynamic field using modified genetic algorithm. *Journal of Computational Science*. https://doi.org/10.1016/j.jocs.2017.08.004.
8. Metawa, N., Hassan, M. K., & Elhoseny, M. (2017). Genetic algorithm based model for optimizing bank lending decisions. *Expert Systems with Applications*, *80*, 75–82. https://doi.org/10.1016/j.eswa.2017.03.021.
9. Elhoseny, M., Shehab, A., & Yuan, X. (2017). Optimizing robot path in dynamic environments using genetic algorithm and bezier curve. *Journal of Intelligent and Fuzzy Systems*, *33*(4), 2305–2316. IOS-Press. https://doi.org/10.3233/JIFS-17348.
10. Elhoseny, M., Elminir, H., Riad, A., & Yuan, X. (2016a). A secure data routing schema for WSN using elliptic curve cryptography and homomorphic encryption. *Journal of King Saud University-Computer and Information Sciences*, *28*(3), 262–275.
11. Elhoseny, M., Yuan, X., El-Minir, H. K., & Riad, A. M. (2016b). An energy efficient encryption method for secure dynamic WSN. *Security and Communication Networks*, *9*(13), 2024–2031.
12. Elsayed, W., Elhoseny, M., Riad, A., & Hassanien, A. E. (2017). Autonomic self-healing approach to eliminate hardware faults in wireless sensor networks. In *International conference on advanced intelligent systems and informatics*, pp. 151–160. Springer.
13. Elsayed, W., Elhoseny, M., Sabbeh, S., & Riad, A. (2017). Self-maintenance model for wireless sensor networks. *Computers and Electrical Engineering*. https://doi.org/10.1016/j.compeleceng.2017.12.022. (In Press).
14. Elhoseny, M., Yuan, X., El-Minir, H. K., & Riad, A. M. (2016). An energy efficient encryption method for secure dynamic WSN. *Security and Communication Networks*, *9*(13), 2024–2031. https://doi.org/10.1002/sec.1459.
15. Hosseinabadi, A. A. R., Vahidi, J., Saemi, B., Sangaiah, A. K., & Elhoseny, M. (2018). Extended genetic algorithm for solving open-shop scheduling problem. *Soft Computing*. https://doi.org/10.1007/s00500-018-3177-y.
16. Elhoseny, M., Ramírez-González, G., Abu-Elnasr, O. M., Shawkat, S. A., Arunkumar, N., & Farouk, A. (2018). Secure medical data transmission model for IoT-based healthcare systems. *IEEE Access*, *PP*(99). https://doi.org/10.1109/ACCESS.2018.2817615.
17. Shehab, A., Elhoseny, M., Muhammad, K., Sangaiah, A. K., Yang, P., Huang, H., & Hou, G. (2018). Secure and robust fragile watermarking scheme for medical images. *IEEE Access*, *6*(1), 10269–10278. https://doi.org/10.1109/ACCESS.2018.2799240.
18. Farouk, A., Batle, J., Elhoseny, M., Naseri, M., Lone, M., Fedorov, A., Alkhambashi, M., Ahmed, S. H. & Abdel-Aty, M., (2018). Robust general N user authentication scheme in a centralized quantum communication network via generalized GHZ states. *Frontiers of Physics*, *13*, 130306. Springer. https://doi.org/10.1007/s11467-017-0717-3.
19. Elhoseny, M., Elkhateb, A., Sahlol, A., & Hassanien, A. E. (2018). Multimodal biometric personal identification and verification. In A. Hassanien, & D. Oliva (Eds.), *Advances in soft computing and machine learning in image processing*. Studies in Computational Intelligence, Vol. 730. Cham: Springer. https://doi.org/10.1007/978-3-319-63754-9_12.
20. Elhoseny, M., Essa, E., Elkhateb, A., Hassanien, A. E., & Hamad, A. (2018). Cascade multimodal biometric system using fingerprint and Iris patterns. In A. Hassanien, K. Shaalan, T.

Gaber, & M. Tolba (Eds.) *Proceedings of the international conference on advanced intelligent systems and informatics 2017*, AISI 2017. Advances in Intelligent Systems and Computing, Vol. 639. Cham: Springer. https://doi.org/10.1007/978-3-319-64861-3_55.

21. Elhoseny, M., Elleithy, K., Elminir, H., Yuan, X., & Riad, A. (2015). Dynamic clustering of heterogeneous wireless sensor networks using a genetic algorithm towards balancing energy exhaustion. *International Journal of Scientific & Engineering Research*, *6*(8), 1243–1252.
22. Elhoseny, M., Hosny, A., Hassanien, A. E., Muhammad, K., & Sangaiah, A. K. (2017). Secure automated forensic investigation for sustainable critical infrastructures compliant with green computing requirements. *IEEE Transactions on Sustainable Computing*, *PP*(99). https://doi.org/10.1109/TSUSC.2017.2782737.
23. Elhoseny, M., Abdelaziz, A., Salama, A. S., Riad, A. M., Muhammad, K., & Sangaiah, A. K. (2018). A hybrid model of internet of things and cloud computing to manage big data in health services applications. *Future Generation Computer Systems*. Elsevier. (In Press).
24. Abdelaziz, A., Elhoseny, M., Salama, A. S., & Riad, A. M. (2018). A machine learning model for improving healthcare services on cloud computing environment. *Measurement119*, 117–128. https://doi.org/10.1016/j.measurement.2018.01.022.
25. Yuan, X., Li, D., Mohapatra, D., & Elhoseny, M. (2017). Automatic removal of complex shadows from indoor videos using transfer learning and dynamic thresholding. *Computers and Electrical Engineering*. https://doi.org/10.1016/j.compeleceng.2017.12.026. (in Press).
26. Sajjad, M., Nasir, M., Muhammad, K., Khan, S., Jan, Z., Sangaiah, A. K., Elhoseny, M., & Baik, S. W. (2017). Raspberry Pi assisted face recognition framework for enhanced law-enforcement services in smart cities. *Future Generation Computer Systems*. Elsevier. https://doi.org/10.1016/j.future.2017.11.013.
27. Shehab, A., Elhoseny, M., El Aziz, M. A., & Hassanien, A. E. (2018). Efficient schemes for playout latency reduction in P2P-VoD systems. In A. Hassanien, & D. Oliva (Eds.), *Advances in soft computing and machine learning in image processing*. Studies in Computational Intelligence, Vol. 730. Cham: Springer. https://doi.org/10.1007/978-3-319-63754-9_22.
28. Elhoseny, M., Nabil, A., Hassanien, A. E., & Oliva, D. (2018). Hybrid rough neural network model for signature recognition. In: A. Hassanien, & D. Oliva (Eds.), *Advances in soft computing and machine learning in image processing*. Studies in Computational Intelligence, Vol. 730. Cham: Springer. https://doi.org/10.1007/978-3-319-63754-9_14.
29. Abdeldaim, A. M., Sahlol, A. T., Elhoseny, M., & Hassanien, A. E. (2018). Computer-aided acute lymphoblastic leukemia diagnosis system based on image analysis. In A. Hassanien, & D. Oliva (Eds.), *Advances in soft computing and machine learning in image processing*. Studies in Computational Intelligence, Vol. 730. Cham: Springer. https://doi.org/10.1007/978-3-319-63754-9.
30. Elhoseny, H., Elhoseny, M., Riad, A. M., & Hassanien, A. E. (2018). A framework for big data analysis in smart cities. In A. Hassanien, M. Tolba, M. Elhoseny, & M. Mostafa (Eds.), *AMLTA 2018 The international conference on advanced machine learning technologies and applications*, (AMLTA2018). Advances in Intelligent Systems and Computing, Vol. 723. Cham: Springer. https://doi.org/10.1007/978-3-319-74690-6_40.
31. Elhoseny, M., Shehab, A., & Osman, L. (2018). An empirical analysis of user behavior for P2P IPTV workloads. In A. Hassanien, M. Tolba, M. Elhoseny, & M. Mostafa (Eds.), *AMLTA 2018 the international conference on advanced machine learning technologies and applications (AMLTA2018)*. Advances in Intelligent Systems and Computing, Vol. 723. Cham: Springer. https://doi.org/10.1007/978-3-319-74690-6_25.
32. Wang, M. M., Qu, Z. G., & Elhoseny, M. (2017). Quantum secret sharing in noisy environment. In X. Sun, H. C. Chao, X. You, & E. Bertino (Eds.), *Cloud Computing and Security*, ICCCS 2017. Lecture Notes in Computer Science, Vol. 10603. Cham: Springer. https://doi.org/10.1007/978-3-319-68542-7_9.
33. Elsayed, W., Elhoseny, M., Riad, A. M., & Hassanien, A. E. (2018). Autonomic self-healing approach to eliminate hardware faults in wireless sensor networks. In A. Hassanien, K. Shaalan, T. Gaber, & M. Tolba (Eds.), *Proceedings of the international conference on advanced intelligent systems and informatics 2017*, AISI 2017. Advances in Intelligent Systems and Computing, Vol. 639. Cham: Springer. https://doi.org/10.1007/978-3-319-64861-3_14.

34. Abdelaziz, A., Elhoseny M., Salama, A. S., Riad, A. M., & Hassanien, A. E. (2018). Intelligent algorithms for optimal selection of virtual machine in cloud environment, towards enhance healthcare services. In A. Hassanien, K. Shaalan, T. Gaber, & M. Tolba (Eds.), *Proceedings of the international conference on advanced intelligent systems and informatics 2017*, AISI 2017. Advances in Intelligent Systems and Computing, Vol. 639. Cham: Springer. https://doi.org/10.1007/978-3-319-64861-3_27.
35. Shehab, A., Ismail, A., Osman, L., Elhoseny, M., & El-Henawy, I. M. (2018). Quantified self using IoT wearable devices. In A. Hassanien, K. Shaalan, T. Gaber, & M. Tolba (Eds.), *Proceedings of the international conference on advanced intelligent systems and informatics 2017*, AISI 2017. Advances in Intelligent Systems and Computing, Vol. 639. Cham: Springer. https://doi.org/10.1007/978-3-319-64861-3_77.
36. Yuan, X., Elhoseny, M., El-Minir, H., & Riad, A. (2017). A genetic algorithm-based, dynamic clustering method towards improved WSN longevity. *Journal of Network and Systems Management*, *25*(1), 21–46.
37. Tharwat, A., Mahdi, H., Elhoseny, M., & Hassanien, A. E. (2018). Recognizing human activity in mobile crowdsensing environment using optimized k-NN algorithm. *Expert Systems With Applications*. https://doi.org/10.1016/j.eswa.2018.04.017. Accessed 12 April 2018.
38. Tharwat, A., Elhoseny, M., Hassanien, A. E., Gabel, T., & Kumar, A. (2018). Intelligent Bezir curve-based path planning model using chaotic particle swarm optimization algorithm. *Cluster Computing*, 1–22. Springer. https://doi.org/10.1007/s10586-018-2360-3.
39. Sarvaghad-Moghaddam, M., Orouji, A. A., Ramezani, Z., Elhoseny, M., & Farouk, A. (2018). Modelling the spice parameters of SOI MOSFET using a combinational algorithm. *Cluster Computing*. Springer. https://doi.org/10.1007/s10586-018-2289-6. (in Press),
40. Rizk-Allah, R. M., Hassanien, A. E., & Elhoseny, M. (2018). A multi-objective transportation model under neutrosophic environment. *Computers and Electrical Engineering*. Elsevier. https://doi.org/10.1016/j.compeleceng.2018.02.024. (in Press)
41. Batle, J., Naseri, M., Ghoranneviss, M., Farouk, A., Alkhambashi, M., & Elhoseny, M. (2017). Shareability of correlations in multiqubit states: Optimization of nonlocal monogamy inequalities. *Physical Review A*, *95*(3):032123. https://doi.org/10.1103/PhysRevA.95.032123.
42. El Aziz, M. A., Hemdan, A. M., Ewees, A. A., Elhoseny, M., Shehab, A., Hassanien, A. E., & Xiong, S. (2017). Prediction of biochar yield using adaptive neuro-fuzzy inference system with particle swarm optimization. In *2017 IEEE PES PowerAfrica conference*, (pp. 115–120), June 27–30, 2017. Accra-Ghana: IEEE. https://doi.org/10.1109/PowerAfrica.2017.7991209.
43. Ewees, A. A., El Aziz, M. A., & Elhoseny, M. (2017). Social-spider optimization algorithm for improving ANFIS to predict biochar yield. In *8th international conference on computing, communication and networking technologies (8ICCCNT)*, July 3–5. Delhi-India: IEEE.
44. Metawa, N., Elhoseny, M., Hassan, M. K., & Hassanien, A. E. (2016). Loan portfolio optimization using genetic algorithm: A case of credit constraints. In *Proceedings of 12th international computer engineering conference (ICENCO)*, pp. 59–64. IEEE. https://doi.org/10.1109/ICENCO.2016.7856446.
45. Ehsan, S., Bradford, K., Brugger, M., Hamdaoui, B., Kovchegov, Y., Johnson, D., et al. (2012). Design and analysis of delay-tolerant sensor networks for monitoring and tracking free-roaming animals. *IEEE Transactions on Wireless Communications*, *11*(3), 1220–1227.
46. Stankovic, J., Wood, D., & Tian, H. (2011). Realistic applications for wireless sensor networks in theoretical aspects of distributed computing in sensor networks. *Monographs in Theoretical Computer Science*, *4*, 835–863.
47. Stankovic, J. (2008). When sensor and actuator networks cover the world. *ETRI Journal*, *30*(5), 627–633.
48. Asim, M., & Mathur, V. (2013). Genetic algorithm based dynamic approach for routing protocols in mobile ad hoc networks. *Journal of Academia and Industrial Research*, *2*(7), 437–441.
49. Raj, E. (2012). An efficient cluster head selection algorithm for wireless sensor networks EDRLEACH. *Journal of Computer Engineering*, *2*(2), 39–44.
50. Ahmed, G., Khan, N., & Ramer, R. (2008). Cluster head selection using evolutionary computing in wireless sensor networks. In *Progress in electromagnetics research symposium*, pp. 883–886.

51. Ramesh, K., & Somasundaram, K. (2011). A comparative study of clusterhead selection algorithms in wireless sensor networks. *International Journal of Computer Science and Engineering Survey*, *2*(4), 153–164.
52. Darwish, A., Hassanien, A. E., Elhoseny, M., Sangaiah, A. K., & Muhammad, K. (2017). The impact of the hybrid platform of internet of things and cloud computing on healthcare systems: Opportunities, challenges, and open problems. *Journal of Ambient Intelligence and Humanized Computing*. https://doi.org/10.1007/s12652-017-0659-1.

Chapter 4
Extending Homogeneous WSN Lifetime in Dynamic Environments Using the Clustering Model

Abstract To extend the longevity of a homogeneous WSN, the key is to avoid nodes deplete energy before the others. Accordingly, this chapter proposes a new clustering model for WSN used in dynamic environments. In each transmission round, the remaining energy of sensor nodes are fairly even with some fluctuations. That is, as a consequence of the proposed method, the variance among remaining energy is quite low, which implies that the sensor nodes shared the burden of relaying messages and, hence, elongated the overall network life. The main factors that are used in our proposed method for choosing a CH are the distance between the CH and BS, the remaining battery power, and the expected consumed energy.

4.1 Hypothesis and Objectives

Despite the great efforts in automatic organizing nodes, the dynamic nature of WSN and numerous possible cluster configurations make searching for an optimal network structure on-the-fly an open challenge [3–5]. In WSNs, sensors are put in open fields without human participation, and gain information coterminously after some time. Limiting the power utilization in vitality constrained sensors is a vital issue to broaden the usefulness and accessibility of a WSN [6, 7]. These limitations represent a huge challenge for most of the recent applications which use a WSN as a main component in their structure [8–21]. To address this need, a novel method for constructing sensor network is proposed with a goal of extending the overall network life. In the proposed method, network clusters are dynamically formed [22, 23]. With the goal of optimizing the lifespan of the entire network, genetic algorithm [24–27] is employed to search for the most suitable sensor nodes as the CHs to relay the messages to the BS. Using the chosen cluster heads, sensor clusters are formed that minimize the total inner cluster node-to-cluster head distance. Compared with state-of-the art methods, the experimental results demonstrated that the proposed method greatly extended the network life. In each transmission round, the remaining energy of sensor nodes are fairly even with some fluctuations. That is, as a consequence of the proposed method, the variance among remaining energy is quite low, which implies that the sensor nodes shared the burden of relaying messages and, hence,

M. Elhoseny and A. E. Hassanien, *Dynamic Wireless Sensor Networks*, Studies in Systems, Decision and Control 165, https://doi.org/10.1007/978-3-319-92807-4_4

elongated the overall network life. The main factors that are used in our proposed method for choosing a CH are the distance between the CH and BS, the remaining battery power, and the expected consumed energy.

Ideally the lifetime of a homogeneous WSN is maximized when the remaining energy of nodes in the network remains the same [28]; that is, no single node completely depletes its energy before the others. This is, however, difficult to achieve in a real-world cluster-based network due to different roles of sensor nodes and various signal transmission distance [29]. The nodes serving as cluster head consume additional energy to fulfill tasks such as receiving messages from member nodes and relaying the aggregated messages to the base station. Balancing node energy consumption and extending the overall network lifespan are non-trivial given many factors that could affect the energy expenditure of each node [30, 87].

Energy consumption of a node is attributed to data acquisition, processing, and transmission [32, 33]. In a complex network, factors such as distance among nodes, distance to the base station, and data throughput are greatly influencing the remaining energy of each node [34–38]. Methods have been developed that account for one or more such factors to achieve extended network longevity. Low Energy Adaptive Clustering Hierarchy (LEACH) [39] is developed to allow different node to share the burden of relaying messages. Rotating the role of cluster heads among nodes in the cluster not only balances the load, but provides a means for fault-tolerance [40]. EEUC [41] selects cluster head based on the distances between the node and the base station. LELE [42] takes remaining energy into consideration and nodes with more energy are more likely to serve as cluster head. In [43], number of neighbors, residual energy, and distance to base station are used in constructing clusters and selecting cluster heads. Without even looking into the characteristics of each single node, the network itself varies greatly in the number of sensor nodes, their placement, and the arrangement of the base station. A pre-determined communication structure or a randomized clustering scheme is far from fulfilling the critical needs of adaptiveness.

4.2 Network Model and Energy Estimation

In this work, the first order radio model (as shown in Fig. 4.1) for a node is adopted. The energy expenditure E of a node s is a summation of energy spent to acquire, process, transmit, and receive data:

$$E_s = E_s^A(l) + E_s^P(l') + E_s^R(l'') + E_s^T(l', d), \tag{4.1}$$

where $E_s^A(l)$ denotes the energy used to acquire l bits of data, $E_s^P(l')$ denotes the energy used to process l' bits of data, $E_s^R(l'')$ denotes the energy used to receive l'' bits of data from nearby nodes if node s serves as a cluster head, and $E_s^T(l', d)$ denotes the energy of transmitting l' bits of data over a distance d. $E_s^R(l'')$ is non-zero only if node s serves as a cluster head. We assume the amount of energy used by each node in data acquisition is the same.

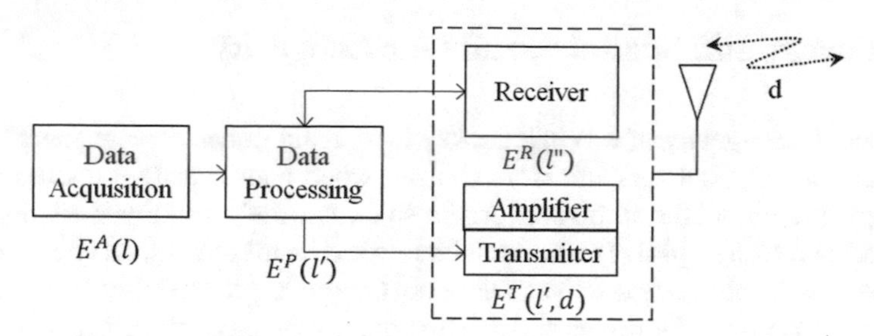

Fig. 4.1 First order radio model of a node. Each component has an energy consumption model that is a function of message length

When computing the energy to transmit or receive a message of l bits, we adopt the following formulas for the transmitter energy consumption E_s^T and receiver energy consumption E_s^R:

$$E_s^T = E_i + ld^n \tag{4.2}$$

$$E_s^R = E_i + lE^* \tag{4.3}$$

where E_i is the idle energy expenditure and d is the distance between the transmitter and receiver as shown in Fig. 4.1. Depending on the distance between the transmitter and receiver, the transmission energy consumption E^T is proportional to different order of the distance and can be modeled by varying the power term n. In this model, we use $n = 4$ for long distance transmission, i.e., transmitting messages from cluster head to base station, and $n = 2$ for short distance transmission, e.g., a node its cluster head. The communications within the cluster take short distance transmission; the communications between the cluster head and base station require long distance transmission. E^* in Eq. (4.3) represents the cost of beam forming approach to reduce the energy consumption.

Based on the consumed energy, the remaining energy of a node s at time t is computed as follows:

$$\tilde{E}_s(t) = E_s(0) - \sum_{i=1}^{t} E_s(i), \tag{4.4}$$

where $E(0)$ is the initial energy of the node and t is the node life time in term of transmission rounds. The remaining energy of every node is updated in each round.

Nodes are stationary and their geospatial locations are known to the base station. In each network transmission round, network structure is computed based on the current network status and broadcast to the nodes. Nodes receive the cluster head assignments and cluster information, and communicate to the base station via the associated cluster head.

4.3 Energy and Spatial Factors for Longevity

To extend the longevity of a WSN, the key is to avoid nodes deplete energy before the others. Ideally, network life is maximized when nodes retain the same amount of energy throughout the entire network lifespan. A widely employed energy metric is the residual energy [44]. The nodes with lower amount of residual energy are less likely to serve as cluster heads due to the extra energy needed to relay messages from the associated nodes. However, the residual energy presents the current status and does not characterize the energy level of a node after the next transmission round, which is a direct measure of the goodness of a network clustering structure. Unless the clustering structure remains the same for the life of the network, the dynamically formed local communication requires different amount of energy contribution from the to-be cluster head. Hence, it is crucial to gauge the expected energy expenditure for each node in a possible network structure.

Assume l bits of data are collected by each node in a round. In a cluster that consists of N_s nodes, the data are aggregated by a node s. Following Eqs. (4.1) and (4.2), the expected energy consumption (denoted with $\hat{E}$) of a non-cluster head node s' and a cluster head node s can be computed as follows:

$$\hat{E}_{s'} = E + lD^2(s', s), \tag{4.5}$$

$$\hat{E}_s = E + N_s l E^* + (N_s + 1) l D^4(s, BS) \tag{4.6}$$

where E is the constant energy expenditure that includes energy used for data acquisition, processing and idle. Functions $D(s', s)$ and $D(s, B)$ give the distances between nodes s' and the cluster head s and from node s to the base station (B), respectively. It is approximated with Euclidean distance.

In addition to the energy metrics, network spatial feature is another influencer to the network life. The spatial characteristics such as distance to the base station and local node density (LSD) provide additional dimensions for intelligent clustering. It becomes more important when the amount of the remaining energy of nodes is very similar, in which case the best network structure is as good as random clustering if spatial factors are not considered.

To characterize local node density, a neighborhood distance δ threshold is used. The density is proportional to the number of neighbors within the δ-vicinity as follows:

$$G_s(\delta) \propto \|S_s\|, \text{and} S_s = \{s_i; D(s, s_i) \leq \delta\} \tag{4.7}$$

where S_s is the set of nodes in the δ-vicinity of s and function $\|\cdot\|$ gives the set size.

4.4 Dynamically Structuring Sensor Network Using GA

Structuring a sensor network requires grouping nearby nodes into clusters and designating a surrogate node as cluster head to each cluster. It is essentially a multi-parameter optimization problem [45–49], in which energy and spatial factors need to be optimized concurrently to achieve overall network longevity. The working steps of GASONeC is shown at Fig. 4.2. First, a random allocation of sensors is created as an input to GASONeC. Based on that, the three working steps are executed in order to get the optimum network configuration. These steps are Cluster Heads Election, Clusters Formation, and Configuration Evaluation. Each of these steps is described in details as the following.

4.4.1 Cluster Head Selection

In clustering model, there are various problems in cluster management process. A major challenge in WSNs is to select appropriate cluster heads in the cluster based model. CHs collect the data from its clusters nodes and forward it to base station. Choosing the CH will affect the network lifetime because of collecting the data from its cluster nodes and forward it to base station require a special characteristics of the CH. Moreover, by rotating the cluster-head, energy consumption is expected to be uniformly distributed among all nodes. Several studies proposed CH selection protocols based on intelligent algorithms as fuzzy logic (FL), genetic algorithm (GA), and neural networks (NNs). But most of CH researches that uses GA depends only on four factors: the distance of a node from the cluster centroid, the remaining battery power, the degree of mobility, and the vulnerability index. While in other researches only two parameters have been considered for number of neighbors and residual energy.

In the CH selection phase of GASONeC, a binary chromosome is used to specify the cluster heads in the network, in which a one represents a cluster head and a

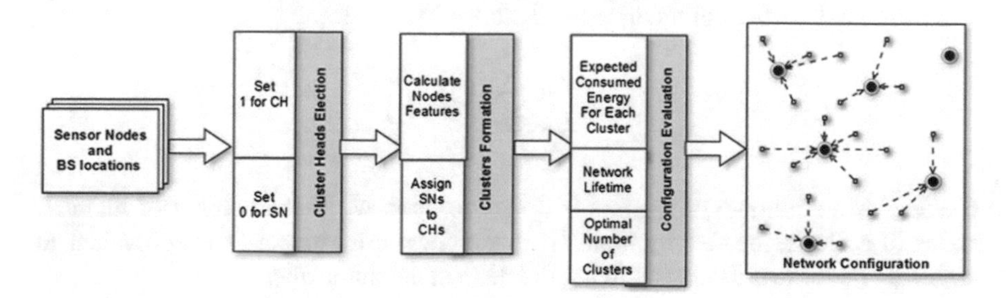

Fig. 4.2 GASONeR consists of three steps: cluster heads election, clusters formation, and configuration evaluation. The arrows depict the data flow

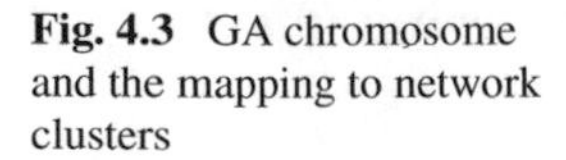

Fig. 4.3 GA chromosome and the mapping to network clusters

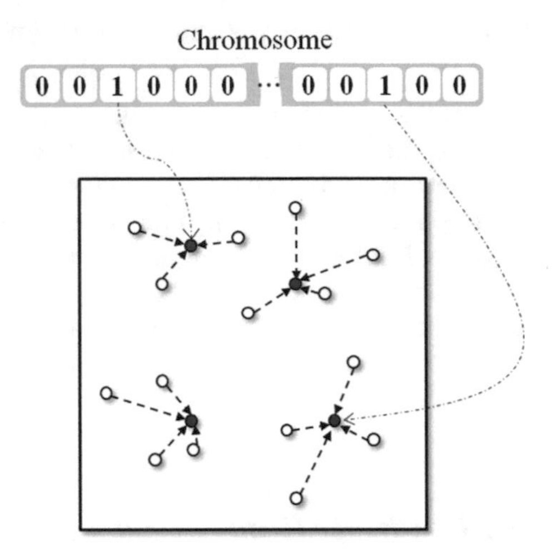

zero represents a member node to a cluster. An example is shown in Fig. 4.3 and the cluster heads are highlighted with filled circles. The dash arrows depict the cluster membership. When a node becomes inactive, i.e., out of power, its corresponding gene value is set to -1, which exempts the node from further GA operations.

4.4.2 Cluster Formation

In GASONeC, the network structure is dynamic. That means, it is changed after each round based on the current nodes characteristics such as their remaining energy. After the GA chooses the CHs, the cluster formation starts to assign each node to one CH following the nearest neighbor rule. The distance between the node and its CH should be less than or equal to the distance threshold θ as described later. In case of long distance between a node and all other CHs, the node will be a CH by itself and connects directly to the BS. The mapping to node clusters from a chromosome is to minimize the network communication distance $\mathbb{D}$ as follows:

$$\mathbb{D} = \sum_{i=1}^{C} \sum_{j=1}^{N_{s_i}} D(s_i, s_j) \tag{4.8}$$

where C is the number of clusters in a network and N_{s_i} is the number of member nodes in a cluster headed by node s_i. In practice, minimizing $\mathbb{D}$ is equivalent to assigning nodes to clusters following the nearest neighbor rule.

4.4.3 Network Configuration Evaluation

Each chromosome represents a network structure. The fitness function is used to evaluate the chromosome by calculating the some of parameters like expected consumed energy of each node. The evaluation process tries to balance the number of clusters to avoid setting each node as a single cluster or grouping all nodes in one cluster. The expected consumed energy for whole network is calculated during the evaluation process as described in the next section.

4.4.4 The Fitness Function and Its Parameters

A key component of GASONeC method is the fitness function. As discussed in Sect. 4.3, there are many factors that influence the network life. Fitness function provides a means to optimize several factors concurrently. The following is a fitness function that integrates energy and spatial factors:

$$f_1 = \sum_s \frac{E_s(t)}{E_s(0)} + \frac{\tilde{E}}{\hat{E}} + \frac{1}{\hat{D}}, \quad (4.9)$$

where $E_s(t)$ is the remaining energy of node s at round t and $E_s(0)$ is the initial energy of node s. $\tilde{E}$ is the total energy cost if the messages are transmitted directly from all nodes to the base station. $\hat{D}$ is the total distance between the cluster heads and the base station:

$$\hat{D} = \sum_{i=1}^{C} D(s_i, BS) \quad (4.10)$$

where each s_i is a node that serves as a cluster head. Without knowledge of the priority of the factors, we assume these three terms are equally important and hence the fitness function takes even weights. However, in cases where it is clear one or more factors play more vital role, uneven weights can be employed in the fitness function. Figure 4.4 summarizes the general operations of GASONeC protocol.

Alternatively, we take into account the local node density as shown in Eq. 4.7 and have the following fitness function:

$$f_2 = \sum_s \frac{E_s(t)}{E_s(0)} + \frac{\tilde{E}}{\hat{E}} + \frac{1}{\hat{D}} + \frac{1}{N} \sum_{s'} G_{s'}(\delta), \quad (4.11)$$

where s' denotes nodes serving as cluster heads. Including node density favors the choice of cluster heads with more close neighbors. Note that the first term on the remaining energy is an aggregation of energy of all nodes including cluster heads and member nodes; whereas the last term includes only the nodes serving as cluster heads.

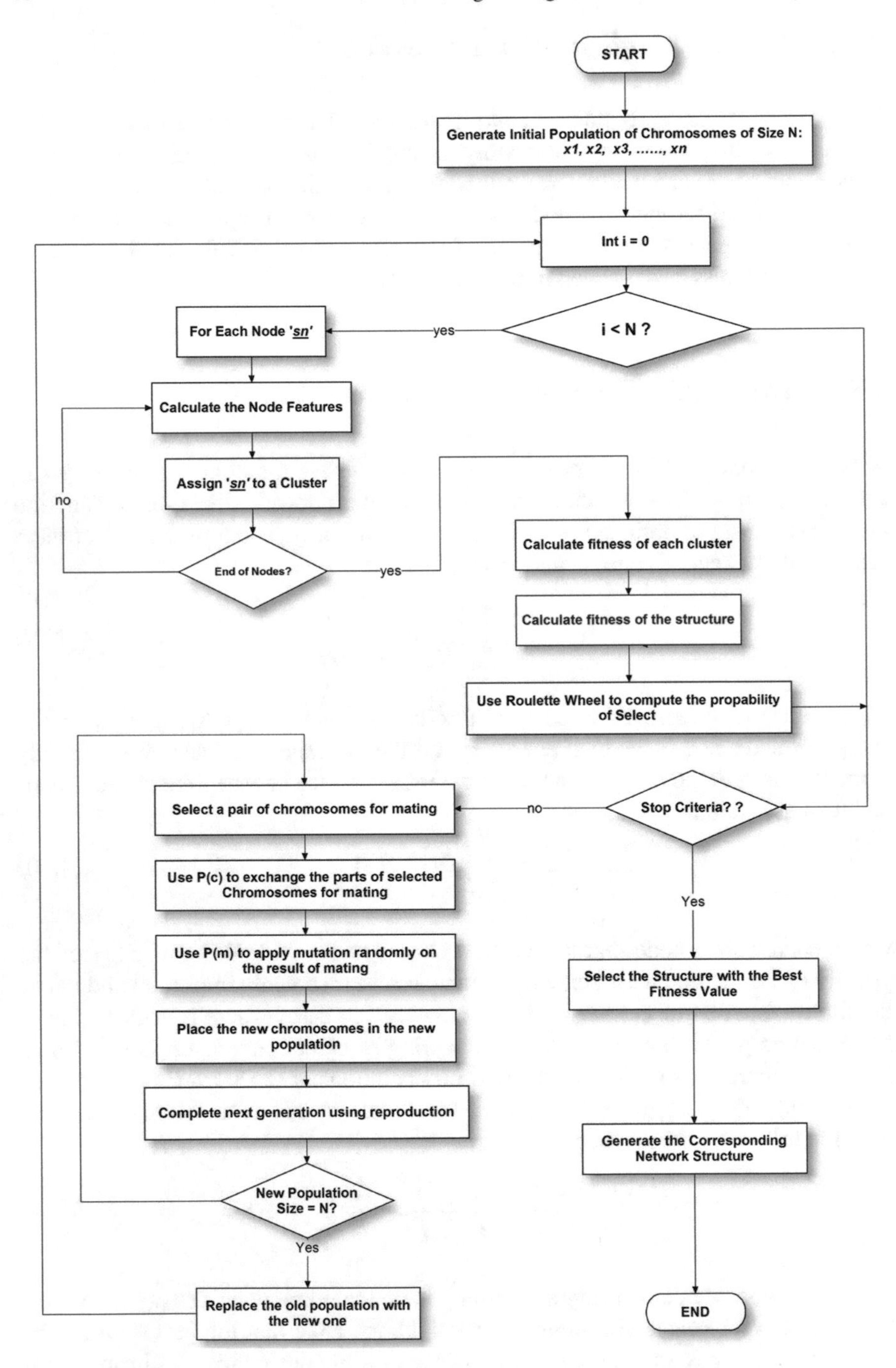

Fig. 4.4 Flowchart that shows the general operations of GASONeC protocol

4.5 Genetic Algorithm-Based, Self-organizing Network Clustering

Algorithm 2 presents GASONeC method. In this algorithm, $q \in [1, Q]$ denotes the number of generations, and the population size is P. The pool of chromosome, denoted by U, is initialized with randomly generated individuals. An intermediate pool of chromosomes, denoted by $\tilde{U}$, is used to hold the individuals created in a generation, and depending on the needs user can specify any intermediate population size that is greater than the initial population size P. In crossover operation, two chromosomes are randomly selected from U and, according to the crossover probability α, two new chromosomes are created by switching consecutive genes. In mutation operations, the value of a randomly picked gene is altered between 0 and 1 according to the mutation probability β. The evolution terminates when one of the following criteria is satisfied: (1) the maximum number of generations is reached; or (2) the fitness converges. Upon completion of the GA evolution, the chromosome that gives the best fitness value is used to restructuring the nodes.

4.6 Results

Table 4.1 lists the network parameters used in these experiments. In running GA, we use the population size of 30 for 30 generations. The crossover probability and mutation probability are 0.8 and 0.006, respectively. The neighborhood distance δ is 20 m throughout these experiments when LSD is calculated. The average performance of 10 repetitions is reported. In each experiment, nodes are randomly placed

Table 4.1 Network properties

Properties		Values
Number of nodes		100
Initial node energy		0.5 J
Idle state energy		50 nJ/bit
Data aggregation energy		5 nJ/bit
Amplification energy	$d \geq d_0$	10 pJ/bit/m^2
(Cluster head to base station)	$d < d_0$	0.0013 pJ/bit/m^2
Amplification energy	$d \geq d_1$	$E_{fs}/10 = E_{fs1}$
(Node to cluster head)	$d < d_1$	$E_{mp}/10 = E_{mp1}$
Packet size		400 bits

Algorithm 2: Genetic Algorithm-Based, Self-organizing Network Clustering.

1: Generate a pool of P chromosomes $U = \{u_1, u_2, \ldots, u_P\}$.
2: $\forall u_i \in U$, structuring the WSN by minimizing Eq. (4.8)
3: Evaluate the fitness of each $u_i \in U$ following Eqs. (4.9) or (4.11).
4: **for** $q = 1, 2, \ldots, Q$ **do**
5: $\quad \tilde{U} \Leftarrow \emptyset$
6: $\quad$ **for** $p = 1, 2, \ldots, P$ **do**
7: $\quad\quad$ Randomly select $u_a, u_b \in U$ $(a \neq b)$ based on the normalized fitness $\tilde{f}(u)$:

$$\tilde{f}(u) = \frac{f(u)}{\sum_c f(u)}.$$

8: $\quad\quad$ Cross over u_a and u_b according to α

$$\mathscr{C}(u_a, u_b|\alpha) \Rightarrow u'_a, u'_b.$$

9: $\quad\quad$ Perform mutation on u'_a and u'_b according to β

$$\mathscr{M}(u'_a|\beta) \Rightarrow \tilde{u}_a, \quad \mathscr{M}(u'_b|\beta) \Rightarrow \tilde{u}_b.$$

10: $\quad\quad$ Evaluate $f(\tilde{u}_a)$ and $f(\tilde{u}_b)$.
11: $\quad\quad$ $\tilde{U} \Leftarrow \tilde{U} \cup \{\tilde{u}_a, \tilde{u}_b\}$
12: $\quad$ **end for**
13: $\quad$ $U \Leftarrow \{u_i; u_i \in \tilde{U}$ and $f(u_i)\}$
14: **end for**
15: Return the chromosome u^* that satisfies

$$u^* = \arg\max_u f(u), u \in U$$

in the field and the base station is also randomly placed at a certain distance to the field center. Comparison studies are conducted with eight state-of-the-art methods including LEACH [39], MODLEACH [50], HEED [44], PEGASIS [51], SEP [52], TSEP [53], M-GEAR [54], DDEEC [55], GA-WCA [56], and GABEEC [57], among which GA-WCA and GABEEC are methods that employ GA.

To maximize the network life, no node shall deplete its energy before others. That is, between transmission rounds, the remaining energy of all nodes is at the same level. In practice, however, this is infeasible, especially when the data volume transmitted within a round is large. Nevertheless, by restructuring the network, the discrepancy among nodes can be suppressed and hence the network life is improved.

It is evident that the remaining energy of all nodes is mostly at the same level throughout the network life. Table 4.2 lists the variance of the remaining energy of these two exemplar cases. The energy discrepancy among nodes increases as the transmission continues. This is inevitable in practice due to the geospatial difference

Table 4.2 Remaining energy (J) variance of nodes. Note the initial energy is 0.5 J

Rounds	Base-station at center			Base-station at boundary		
	500	1000	1500	500	1000	1500
Mean	0.393	0.285	0.179	0.381	0.261	0.141
STD	0.003	0.059	0.088	0.026	0.053	0.078

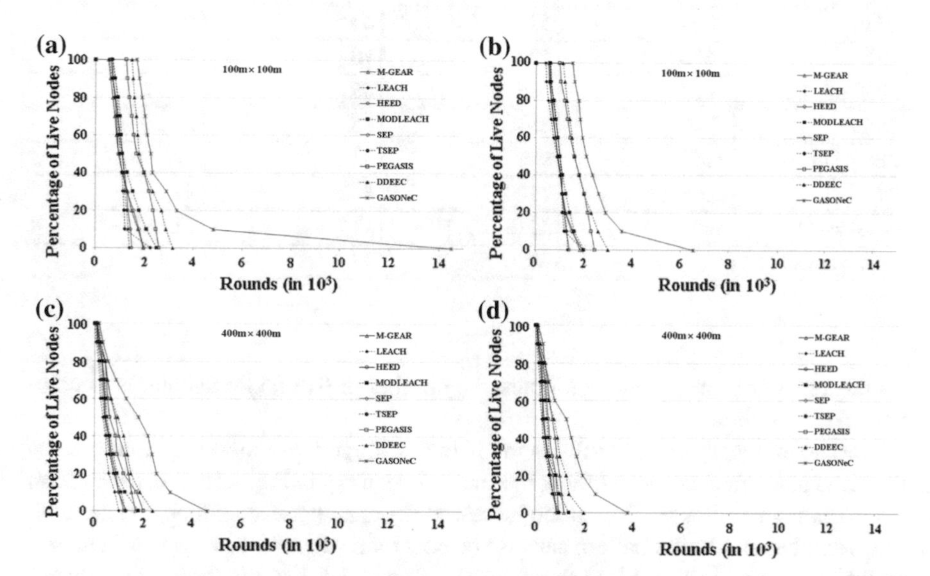

Fig. 4.5 Network lifetime in terms of network transmission rounds. **a** base station in the center of a 100 m × 100 m field. For clear visualization and consistence among all figures, the x-axis in (**a**) is truncated. The number of round when the last node depletes all its energy for GASONeC is 14500. **b** base station on the boundary of a 100 m × 100 m field. **c** base station in the center of a 400 m × 400 m field. **d** base station on the boundary of a 400 m × 400 m field

among nodes. On the other hand, if we allow more generations in GA, it is possible to identify better network structure to minimize the variance of the remaining energy. As a consequence, it takes longer time to come to a solution, which impacts the efficiency of the network. Figure 4.5 depicts the average number of live nodes throughout the entire network lifespan using the state-of-the-art methods. In Fig. 4.5a and b, 100 nodes are randomly placed in a 100 m× 100 m field. Figure 4.5a illustrates results with the base station placed in the field center, and Fig. 4.5b illustrates the results with the base station placed on the field boundary. Figure 4.5c and d illustrate results of a sensor field with randomly placed nodes in an area of 400 m × 400 m. The base station locations for Fig. 4.5c and d are at the field center and on the field boundary, respectively. In these plots, the x-axis shows the number of network transmission rounds (in thousands); whereas the y-axis shows the percentage of live nodes. Note that the number of nodes deployed in the field is the same in all cases (see Table 4.1).

Table 4.3 The average number of network transmission rounds before the first node became unavailable due to energy exhaustion

Methods	100 m × 100 m		400 m × 400 m	
	Center	Boundary	Center	Boundary
M-GEAR	510	530	102	80
LEACH	528	390	10	7
HEED	603	570	65	50
SEP	672	500	70	54
MODLEACH	670	586	50	43
TSEP	710	585	35	33
PEGASIS	1255	980	91	78
DDEEC	1496	1100	122	92
GASONeC	**1691**	**1515**	**175**	**112**
Improvement (%)	13	37.3	43.4	21.7

As the network transmission continues, the number of live nodes decreases as more nodes deplete their energy.

The node density greatly affects the network longevity. Figure 4.5a and c show the same placement of base station, but the node density in Fig. 4.5c, i.e., field 400 m × 400 m, is much lower. Due to the increased distance between nodes, the network life is shortened. Similar pattern can also be observed when the base station is placed on the boundary of the field, as shown in Fig. 4.5b and d. Essentially, the density is a transformed view of distance. As the distance between nodes increases, the advantage of forming node clusters becomes less significant in extending the network life.

Placement of base station also has an impact to the network life, particularly to GASONeC method. Comparing Fig. 4.5a and b, the network life using GASONeC is much longer when the base station is placed in the center of the field. The other methods are not affected as much as GASONeC although decrease of network life is observed with the base station placed on the field boundary. Table 4.3 presents the average network transmission round of different methods when the first node became unavailable due to energy exhaustion. By comparing the results of base station placement, it is clear that as the distance between nodes and the base station increases, the average network life is shortened. This is mostly due to the extra energy required to forward data to the base station. If we compare the results from different fields, the networks in a larger area (in this case, it translates to sparse sensor network) tend to have shorter lifespan. In these experiments, GASONeC facilitated the greatest number of network transmission rounds. In contrast to the second best results, the maximum improvement rate of network life is 43.4%.

Communication to the cluster head in a dense sensor area usually costs less energy. Hence, selecting a node that has more close neighboring nodes to serve as a cluster head is advantageous to reduce energy expenditure. In GASONeC method, we devise a fitness function that includes the energy terms as well as local node density to

Table 4.4 Network life span with and without using the number of neighbors in the fitness function. FND: round at which first node die. LND: round at which last node die

Sensor field	w/o local node density		w. local node density	
	FND	LND	FND	LND
100 m × 100 m	1943	12688	7752	15034
400 m × 400 m	175	4633	150	4610

facilitate joint optimization. It is clear that by including the local node density in the GA fitness function, the network life is significantly extended. The network transmission rounds at which the first and the last nodes becoming unavailable are reported in Table 4.4. The network life improvement is about 4 folds with respect to the round of first node discontinuation.

4.6.1 Impact of the Base-Station Location to Network Structure and Life

Intuitively, when the base station is placed far from the sensor field, it is preferred that clusters are formed to conserve energy. However, this contradicts to maximizing the network life because a significant amount of energy is usually required by a cluster head to relay messages from its member nodes. Table 4.5 lists the average number of clusters in a sensor field of 100 m× 100 m. It is clear that including local node density in the GA fitness function makes little difference in the number of clusters. In all cases, the cluster counts are very close. Despite the number of clusters increases when the number of node increases, the proportion of cluster count and node count remains close. As the distance of base station enlarges, the number of clusters increases. With the goal of minimizing the energy discrepancy among nodes, small clusters are

Table 4.5 The average number of clusters in a sensor field of 100 m × 100 m. The distance of base station is with respect to the center of the field

Base-station distance (m)	100 nodes		50 nodes	
	w/o LSD	w. LSD	w/o LSD	w. LSD
300	58	60	29	30
250	51	49	18	21
200	28	32	14	16
150	18	20	10	10
100	11	12	6	7
0	6	8	4	4

Table 4.6 Average network life with respect to the base station locations. The sensor field size is 100 m × 100 m

Base-station	Distance (m)	Without LSD		With LSD	
		FND	LND	FND	LND
50 nodes	300	269	1927	110	2429
	250	646	2086	315	3860
	200	875	2700	1169	2229
	150	1199	5851	3304	8326
	100	1682	10434	4986	11566
	0	2005	12206	7838	16448
100 nodes	300	317	2139	231	2663
	250	628	2486	588	4307
	200	870	2875	1429	2778
	150	1412	5661	3507	8681
	100	1796	9748	5084	12818
	0	2044	13624	8549	18411

favored by GASONeC when the base station is far from the field. Table 4.6 presents the average network life with respect to the base station distance to the sensor field. Both fitness functions with and without local node density are considered in this comparison. Clearly the network life reduced when the base station is placed farther away from the sensor field. It is, however, interesting to observe that as the number of node increased in the field, the network life is slightly improved. Despite that the volume of data acquired and transmitted within the network increased, the larger number of nodes helped to subside the demands of long-distance data transmission, which facilitated extended network life. Also in this 100 m × 100 m field, local node density also served as a factor to extend network life when the base station is relatively closer. As shown in the table, when the distance of the base station to the field center is greater than 200 m, the network life (with respect to the first node die) based on fitness function with local node density greatly decreased. Otherwise, using local node density improved network life. Figure 4.6 illustrates the average network life (in terms of network transmission rounds) with respect to the node density. Figure 4.6a and b depict the results with 50 nodes in the 100 m × 100 m field, and Fig. 4.6c and d depict the results with 100 nodes in the same field. It is evident that the location of base station dramatically affects the network life. When the base station is placed far away from the sensor field, more energy is consumed to transmit data and hence the network life is shortened. Comparing the network life change between base station placed at 0 (i.e., center of the field) and 100 (or 100 and 150, etc.) shown in Fig. 4.6b and that shown in Fig. 4.6a, we can observe that such impact is more significant when local node density is employed in the optimization. Similar observations can be obtained from Fig. 4.6c and d.

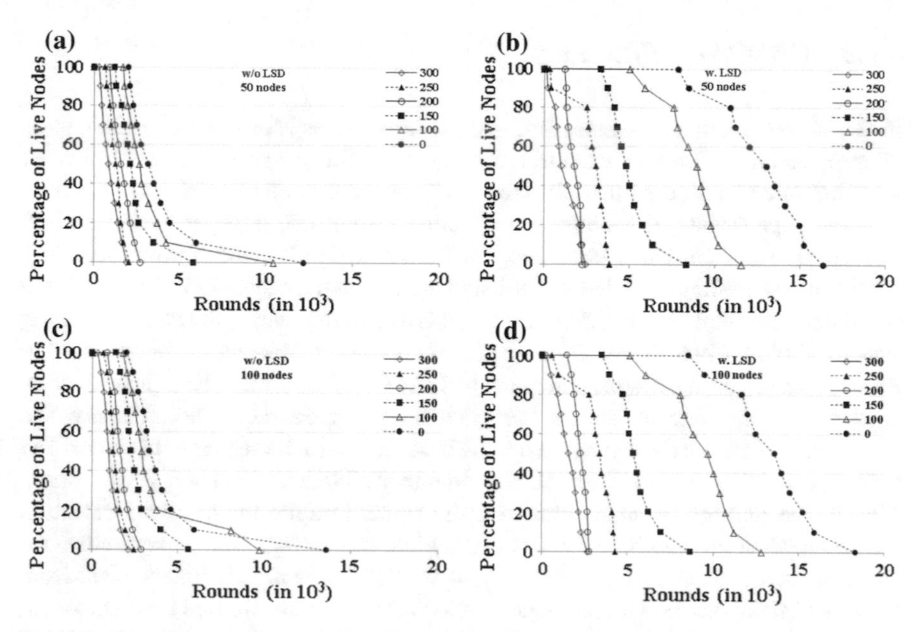

Fig. 4.6 Network life in a sensor field of 100 m × 100 m. **a** 50 nodes in the field and the fitness function without local node density. **b** 50 nodes in the field and the fitness function with local node density. **c** 100 nodes in the field and the fitness function without local node density. **d** 100 nodes in the field and the fitness function with local node density

Table 4.7 Average time (in seconds) used to identify optimal network structure in each transmission round using GASONeC

Field size	100 m × 100 m				400 m × 400 m	
Node count	50		100		100	
BS to field	0 m	100 m	0 m	100 m	0 m	400 m
Average time	0.51	0.57	0.66	0.74	0.59	0.57
Standard dev.	0.05	0.05	0.39	0.16	0.36	0.32

It is interesting to note that when local node density is used in the optimization, the network life is not extended inversely proportional to the base station distance. The exceptional cases are shown in Fig. 4.6b and d when base station is placed at 200 and 250 m. The network life is surprisingly longer when the base station is placed at 250 m compared to that of when the base station is placed at 200 m. Note that the location of base station is randomized on the radius of certain distances to the sensor field center and the network life reported is the average of 10 independent runs. This is possibly due to the instability of including local node density in the fitness function. In contrast, such anomaly is absent in Fig. 4.6a and c.

4.6.2 GASONeC Efficiency

In GASONeC method, the most time consuming process is evaluation of the fitness of chromosomes. Given a chromosome, clusters are formed according to the nearest neighbor rule and fitness function is then evaluated based on this network structure. In these experiments, GA employed 30 chromosomes in any generation and 60 offspring are created. Thirty generations are performed to conclude the optimization. The methods are implemented in C# language and experiments are conducted in a computer with Intel core i5 2.6 GHz CPU, 4 GB memory, and Windows 7 operating system. Table 4.7 lists the average time (in seconds) and standard deviation used to form clusters in each transmission round by GASONeC method. The fitness function used in these experiments includes energy terms, distance and local node density. The "BS to field" denotes the distance from the base station to the sensor field center. The time reported is before the first node became unavailable due to energy exhaustion. That is, the number of nodes remained the same. Despite the standard deviation increased when the number of nodes is doubled, the average time is very close for all cases. It is evident that the efficiency of GASONeC is mostly independent from the sensor field size and the number of nodes. The overall average time across all experiments is 0.58 s with a standard deviation of 0.05. The efficiency of GASONeC is satisfactory.

4.7 Conclusion

A proposed protocol for dynamic clustering homogeneous WSN using genetic algorithm (GA) was explained. The phased towards building this protocols are defined as three main phases: design, testing, and performance evaluation. In addition, the proposed energy model as well as the main GA operations to construct the network structure were discussed. The main steps to construct the network structure are cluster head selection, cluster formation, and configuration evaluation. However, the proposed model assumes that the network is secure and no additional security procedures [58–62] are needed.

References

1. Sarvaghad-Moghaddam, M., Orouji, A. A., Ramezani, Z., Elhoseny, M., & Farouk, A. (2018). Modelling the spice parameters of SOI MOSFET using a combinational algorithm. *Cluster Computing*. Springer. https://doi.org/10.1007/s10586-018-2289-6. (in Press).
2. Wu, Y., & Liu, W. (2013). Routing protocol based on genetic algorithm for energy harvesting-wireless sensor networks. *IET Wireless Sensor Systems*, *3*(2), 112–118.
3. Elhoseny, M., Farouk, A., Zhou, N., Wang, M., Abdalla, S., & Batle, J. (2017a). Dynamic multi-hop clustering in a wireless sensor network: Performance improvement. *Wireless Personal Communications*, 1–21.

4. Tharwat, A., Mahdi, H., Elhoseny, M., & Hassanien, A. E. (2018). Recognizing human activity in mobile crowdsensing environment using optimized k-NN algorithm. *Expert Systems With Applications*. https://doi.org/10.1016/j.eswa.2018.04.017. Accessed 12 April 2018.
5. Tharwat, A., Elhoseny, M., Hassanien, A. E., Gabel, T., & Kumar, A. (2018). Intelligent Bezir curve-based path planning model using chaotic particle swarm optimization algorithm. *Cluster Computing*, 1–22. Springer. https://doi.org/10.1007/s10586-018-2360-3.
6. Elhoseny, M., Tharwat, A., Farouk, A., & Hassanien, A. E. (2017b). K-coverage model based on genetic algorithm to extend WSN lifetime. *IEEE Sensors Letters*, *1*(4), 1–4.
7. Elhoseny, M., Tharwat, A., Yuan, X., & Hassanien, A. E. (2018). Optimizing K-coverage of mobile WSNs, *Expert Systems with Applications*, *92*, 142–153. https://doi.org/10.1016/j.eswa.2017.09.008.
8. Elhoseny, M., Abdelaziz, A., Salama, A. S., Riad, A. M., Muhammad, K., & Sangaiah, A. K. (2018). A hybrid model of internet of things and cloud computing to manage big data in health services applications. *Future Generation Computer Systems*. Elsevier. (in Press).
9. Abdelaziz, A., Elhoseny, M., Salama, A. S., & Riad, A. M. (2018). A machine learning model for improving healthcare services on cloud computing environment. *Measurement, 119*, 117–128. https://doi.org/10.1016/j.measurement.2018.01.022.
10. Darwish, A., Hassanien, A. E., Elhoseny, M., Sangaiah, A. K., & Muhammad, K. (2017). The impact of the hybrid platform of internet of things and cloud computing on healthcare systems: Opportunities, challenges, and open problems. *Journal of Ambient Intelligence and Humanized Computing*. Springer. https://doi.org/10.1007/s12652-017-0659-1.
11. Darwish, A., Hassanien, A. E., Elhoseny, M., Sangaiah, A. K., & Muhammad, K. (2017). Automatic removal of complex shadows from indoor videos using transfer learning and dynamic thresholding. *Computers and Electrical Engineering*. https://doi.org/10.1016/j.compeleceng.2017.12.026. (in Press).
12. Sajjad, M., Nasir, M., Muhammad, K., Khan, S., Jan, Z., Sangaiah, A.K., Elhoseny, M., & Baik, S.W., (2017). Raspberry Pi assisted face recognition framework for enhanced law-enforcement services in smart cities. *Future Generation Computer Systems*. Elsevier. https://doi.org/10.1016/j.future.2017.11.013.
13. Shehab, A., Elhoseny M., El Aziz, M. A., & Hassanien, A. E. (2018). Efficient schemes for playout latency reduction in P2P-VoD systems. In A. Hassanien, & D. Oliva (Eds.), *Advances in soft computing and machine learning in image processing*. Studies in Computational Intelligence, (Vol. 730). Cham: Springer. https://doi.org/10.1007/978-3-319-63754-9_22.
14. Elhoseny M., Nabil A., Hassanien A. E., & Oliva D. (2018). Hybrid rough neural network model for signature recognition. In A. Hassanien, & D. Oliva (Eds.), *Advances in soft computing and machine learning in image processing*. Studies in Computational Intelligence, Vol. 730. Cham: Springer. https://doi.org/10.1007/978-3-319-63754-9_14.
15. Abdeldaim, A. M., Sahlol, A. T., Elhoseny, M., & Hassanien, A. E. (2018). Computer-aided acute lymphoblastic Leukemia diagnosis system based on image analysis. In: Hassanien A., & Oliva D. (eds) *Advances in soft computing and machine learning in image processing*. Studies in Computational Intelligence, Vol. 730. Cham: Springer. https://doi.org/10.1007/978-3-319-63754-9.
16. Elhoseny, H., Elhoseny, M., Riad, A. M., & Hassanien, A. E. (2018). A framework for big data analysis in smart cities. In A. Hassanien, M. Tolba, M. Elhoseny, & M. Mostafa (Eds.) *AMLTA 2018 the international conference on advanced machine learning technologies and applications* (AMLTA2018). Advances in Intelligent Systems and Computing, Vol. 723. Cham: Springer, https://doi.org/10.1007/978-3-319-74690-6_40.
17. Elhoseny, M., Shehab, A., & Osman, L. (2018). An empirical analysis of user behavior for P2P IPTV workloads. In A. Hassanien, M. Tolba, M. Elhoseny, & M. Mostafa (Eds.), *AMLTA 2018 the international conference on advanced machine learning technologies and applications* (AMLTA2018). Advances in Intelligent Systems and Computing, Vol. 723. Cham: Springer. https://doi.org/10.1007/978-3-319-74690-6_25.
18. Wang, M. M., Qu, Z. G., Elhoseny, M. (2017). Quantum secret sharing in noisy environment. In X. Sun, H. C. Chao, X. You, & E. Bertino (Eds.), *Cloud computing and security*, ICCCS 2017.

Lecture Notes in Computer Science, Vol. 10603. Cham: Springer. https://doi.org/10.1007/978-3-319-68542-7_9.
19. Elsayed, W., Elhoseny, M., Riad, A. M., & Hassanien, A. E. (2018). Autonomic self-healing approach to eliminate hardware faults in wireless sensor networks. In A. Hassanien, K. Shaalan, T. Gaber, & M. Tolba (Eds.), *Proceedings of the international conference on advanced intelligent systems and informatics 2017*, AISI 2017. Advances in Intelligent Systems and Computing, Vol. 639. Cham: Springer. https://doi.org/10.1007/978-3-319-64861-3_14.
20. Abdelaziz, A., Elhoseny, M., Salama, A. S., Riad, A. M., & Hassanien, A. E. (2018). Intelligent algorithms for optimal selection of virtual machine in cloud environment, towards enhance healthcare services. In A. Hassanien, K. Shaalan, T. Gaber, & M. Tolba (Eds.), *Proceedings of the international conference on advanced intelligent systems and informatics 2017*, AISI 2017. Advances in Intelligent Systems and Computing, Vol. 639. Cham: Springer. https://doi.org/10.1007/978-3-319-64861-3_27.
21. Shehab, A., Ismail, A., Osman, L., Elhoseny, M., & El-Henawy, I. M. (2018). Quantified self using IoT wearable devices. In A. Hassanien, K. Shaalan, T. Gaber, & M. Tolba (Eds.), *Proceedings of the international conference on advanced intelligent systems and informatics 2017*, AISI 2017. Advances in Intelligent Systems and Computing, Vol. 639. Cham: Springer. https://doi.org/10.1007/978-3-319-64861-3_77.
22. Elhoseny, M., Elleithy, K., Elminir, H., Yuan, X., & Riad, A. (2015). Dynamic clustering of heterogeneous wireless sensor networks using a genetic algorithm towards balancing energy exhaustion. *International Journal of Scientific & Engineering Research*, *6*(8), 1243–1252.
23. Yuan, X., Elhoseny, M., El-Minir, H., & Riad, A. (2017). A genetic algorithm-based, dynamic clustering method towards improved WSN longevity. *Journal of Network and Systems Management*, *25*(1), 21–46.
24. Elhoseny, M., Tharwat, A., & Hassanien, A. E. (2017c). Bezier curve based path planning in a dynamic field using modified genetic algorithm. *Journal of Computational Science*. https://doi.org/10.1016/j.jocs.2017.08.004.
25. Metawa, N., Hassan, M. K., & Elhoseny, M. (2017). Genetic algorithm based model for optimizing bank lending decisions. *Expert Systems with Applications*, *80*, 75–82. https://doi.org/10.1016/j.eswa.2017.03.021.
26. Elhoseny, M., Shehab, A., & Yuan, X. (2017). Optimizing robot path in dynamic environments using genetic algorithm and Bezier curve. *Journal of Intelligent and Fuzzy Systems*, 33(4), 2305–2316. IOS-Press. https://doi.org/10.3233/JIFS-17348.
27. Hosseinabadi, A. A. R., Vahidi, J., Saemi, B., Sangaiah, A. K., & Elhoseny, M. (2018). Extended genetic algorithm for solving open-shop scheduling problem. *Soft Computing*. https://doi.org/10.1007/s00500-018-3177-y.
28. Elhoseny, M., Farouk, A., Batle, J., Shehab, A., & Hassanien, A. E. (2017). Secure image processing and transmission schema in cluster-based wireless sensor network. In *Handbook of research on machine learning innovations and trends*, (Chapter 45, pp. 1022–1040), IGI Global, 2017. https://doi.org/10.4018/978-1-5225-2229-4.ch045.
29. Elhoseny, M., Hosny, A., Hassanien, A. E., Muhammad, K., & Sangaiah, A. K. (2017). Secure automated forensic investigation for sustainable critical infrastructures compliant with green computing requirements. *IEEE Transactions on Sustainable Computing*, *PP*(99). https://doi.org/10.1109/TSUSC.2017.2782737.
30. Tripathi, K., Singh, N., & Verma, K. (2012). Two-tiered wireless sensor networks–base station optimal positioning case study. *IET Wireless Sensor Systems*, *2*(4), 351–360.
31. Wang, L., Wang, C., & Liu, C. (2009). Optimal number of clusters in dense wireless sensor networks: A cross-layer approach. *IEEE Transactions on Vehicular Technology*, *58*(2), 966–976.
32. Elhoseny, M., Yuan, X., El-Minir, H. K., & Riad, A. (2014). Extending self-organizing network availability using genetic algorithm. In *International conference on computing, communication and networking technologies (ICCCNT)*, (pp. 1–6). IEEE.
33. Elhoseny, M., Yuan, X., Yu, Z., Mao, C., El-Minir, H., & Riad, A. (2015). Balancing energy consumption in heterogeneous wireless sensor networks using genetic algorithm. *IEEE Communications Letters*, *19*(12), 2194–2197.

34. Elhoseny, M., Elminir, H., Riad, A., & Yuan, X. (2016a). A secure data routing schema for wsn using elliptic curve cryptography and homomorphic encryption. *Journal of King Saud University-Computer and Information Sciences*, *28*(3), 262–275.
35. Elhoseny, M., Yuan, X., El-Minir, H. K., & Riad, A. M. (2016b). An energy efficient encryption method for secure dynamic WSN. *Security and Communication Networks*, *9*(13), 2024–2031.
36. Elsayed, W., Elhoseny, M., Riad, A., & Hassanien, A. E. (2017). Autonomic self-healing approach to eliminate hardware faults in wireless sensor networks. In *International conference on advanced intelligent systems and informatics*, pp. 151–160. Springer.
37. Elsayed, W., Elhoseny, M., Sabbeh, S., & Riad, A. (2017). Self-maintenance model for wireless sensor networks. *Computers and Electrical Engineering*. https://doi.org/10.1016/j.compeleceng.2017.12.022. (in Press).
38. Elhoseny, M., Yuan, X., ElMinir, H. K., & Riad, A. M. (2016). An energy efficient encryption method for secure dynamic WSN. *Security and Communication Networks*, *9*(13), 2024–2031. https://doi.org/10.1002/sec.1459.
39. Heinzelman, W., Chandrakasan, A., & Balakrishnan, H. (2000). Energy-efficient communication protocol for wireless microsensor networks. In *The Hawaii international conference on system sciences*, Maui, Hawaii.
40. Heinzelman, W., Chandrakasan, A., & Balakrishnan, H. (2002). An application-specific protocol architecture for wireless microsensor networks. *IEEE Transaction Wireless Communications*, *1*(4), 660–670.
41. Chengfa, L., Mao, Y., Guihai, C., & Lie, W. (2005). An energy-efficient unequal clustering mechanism for wireless sensor networks. In *IEEE international conference on mobile Ad hoc and sensor systems*, Washington, DC.
42. Shirmohammadi, M., Faez, K., & Chhardoli, M. (2009). LELE: Leader election with load balancing energy. In *International conference on communications and mobile computing*, (pp. 106–110).
43. Raj, E. (2012). An efficient cluster head selection algorithm for wireless sensor networks EDRLEACH. *Journal of Computer Engineering*, *2*(2), 39–44.
44. Younis, O., & Fahmy, S. (2004). HEED: A hybrid, energy-efficient, distributed clustering approach for ad hoc sensor networks. *IEEE Transactions on Mobile Computing*, *3*(4), 366–379.
45. Rizk-Allah, R. M., Hassanien, A. E., & Elhoseny, M. (2018). A multi-objective transportation model under neutrosophic environment. *Computers and Electrical Engineering*. Elsevier. https://doi.org/10.1016/j.compeleceng.2018.02.024. (in Press).
46. Batle, J., Naseri, M., Ghoranneviss, M., Farouk, A., Alkhambashi, M., & Elhoseny, M. (2017). Shareability of correlations in multiqubit states: Optimization of nonlocal monogamy inequalities. *Physical Review A*, *95*(3), 032123. https://doi.org/10.1103/PhysRevA.95.032123.
47. El Aziz, M. A., Hemdan, A. M., Ewees, A. A., Elhoseny, M., Shehab, A., Hassanien, A. E., & Xiong, S. (2017). Prediction of biochar yield using adaptive neuro-fuzzy inference system with particle swarm optimization. In *2017 IEEE PES PowerAfrica Conference*, (pp. 115–120), June 27–30, 2017. Accra-Ghana: IEEE. https://doi.org/10.1109/PowerAfrica.2017.7991209.
48. Ewees, A. A., El Aziz, M. A., & Elhoseny, M. (2017). Social-spider optimization algorithm for improving ANFIS to predict biochar yield. In *8th International conference on computing, communication and networking technologies* (8ICCCNT), July 3—5. Delhi-India: IEEE.
49. Metawa, N., Elhoseny, M., Hassan, M. K., & Hassanien, A. E. (2016). Loan portfolio optimization using genetic algorithm: A case of credit constraints. In *Proceedings of 12th international computer engineering conference (ICENCO)*, (pp. 59–64). IEEE. https://doi.org/10.1109/ICENCO.2016.7856446.
50. Mahmood, D., Javaid, N., Mahmood, S., Qureshi, S., Memon, A., & Zaman, T. (2013). MOD-LEACH a variant of LEACH for WSNs. In *Eighth international conference on broadband and wireless computing and communication and applications*, (pp. 158–163).
51. Lindsey, S., & Raghavendra, C. (2002). Pegasis power-efficient gathering in sensor information systems. *IEEE Aerospace Conference Proceedings*, *3*, 1125–1130.

52. Smaragdakis, G., Matta, I., & Bestavros, A. (2004). SEP: A stable election protocol for clustered heterogeneous wireless sensor network. In *Second international workshop on sensor and actor network protocols and applications*.
53. Kashaf, A., Javaid, N., Khan, Z., & Khan, I. (2012). TSEP: Threshold-sensitive stable election protocol for WSNs. In *Conference on Frontiers of information technology*, (pp. 164–168).
54. Nadeem, Q., Rasheed, M., Javaid1, N., Khan, Z., Maqsood, Y., & Din, A. (2013). M-GEAR gateway-based energy-aware multi-hop routing protocol for WSNs. In *Eighth international conference on broadband and wireless computing and communication and applications*, (pp. 164–169).
55. Elbhiri, B., Rachid, S., & Elfkihi, S. (2010). Developed distributed energy-efficient clustering (DDEEC) for heterogeneous wireless sensor. In *Communications and mobile network*, (pp. 1–4). Rabat.
56. Nandi, B., Barman, S., & Paul, S. (2010). Genetic algorithm based optimization of clustering in ad-hoc networks. *International Journal of Computer Science and Information Security*, *7*(1), 165–169.
57. Bayrakl, S., & Erdogan, S. (2012). Genetic algorithm based energy efficient clusters in wireless sensor networks. *Procedia Computer Science*, *10*, 247–254.
58. Elhoseny, M., Ramírez-González, G., Abu-Elnasr, O. M., Shawkat, S. A., Arunkumar, N., & Farouk, A. (2018). Secure medical data transmission model for IoT-based healthcare systems. *IEEE Access*, *PP*(99). DOIurlhttps://doi.org/10.1109/ACCESS.2018.2817615.
59. Rizk-Allah, R. M., Hassanien, A. E., & Elhoseny, M. (2018). Secure and robust fragile watermarking scheme for medical images. *IEEE Access*, *6*(1), 10269–10278. https://doi.org/10.1109/ACCESS.2018.2799240.
60. Farouk, A., Batle, J., Elhoseny, M., Naseri, M., Lone, M., Fedorov, A., Alkhambashi, M., Ahmed, S. H., & Abdel-Aty, M. (2018). Robust general N user authentication scheme in a centralized quantum communication network via generalized GHZ states. *Frontiers of Physics*, *13*, 130306. Springer. https://doi.org/10.1007/s11467-017-0717-3.
61. Elhoseny, M., Elkhateb, A., Sahlol, A., & Hassanien, A. E. (2018). Multimodal biometric personal identification and verification. In A. Hassanien, & D. Oliva (Eds.) *Advances in soft computing and machine learning in image processing*. Studies in Computational Intelligence, Vol. 730. Cham: Springer. https://doi.org/10.1007/978-3-319-63754-9_12.
62. Elhoseny, M., Essa, E., Elkhateb, A., Hassanien, A. E., & Hamad, A. (2018). Cascade multimodal biometric system using fingerprint and Iris patterns. In A. Hassanien, K. Shaalan, T. Gaber, & M. Tolba (Eds.), *Proceedings of the international conference on advanced intelligent systems and informatics 2017*, AISI 2017. Advances in Intelligent Systems and Computing, Vol. 639. Cham: Springer. https://doi.org/10.1007/978-3-319-64861-3_55.

Chapter 5
Optimizing Cluster Head Selection in WSN to Prolong Its Existence

Abstract In wireless sensor networks (WNSs), the amount of transferred data is mainly depending on the network lifetime. Hence, the network throughput can be maximized by extending the network lifetime as long as possible. Accordingly, the clustering model is proposed to extend the network lifetime and improve the network performance. However, the optimum network structure in that model may differs from round to round depending on a set of sensor nodes characteristics, i.e, their remaining energy. Getting the intended optimum structure is non trivial process, which includes determining the appropriate number of clusters, electing a cluster head (CH) for each cluster, and assigning each sensor node to a clusters. For that, a new Genetic Algorithm (GA) based model is proposed to form the network structure that optimize its throughput.

5.1 Introduction

Currently, Wireless sensor network (WSNs) are used in many environments, such as smart cities applications, for monitoring and tracking. Usually, all sensors are provided by one-time use batteries which make saving energy an urgent need [1–8]. Hence, the management of sensors is an important factor for extending the availability of a WSN [9]. An optimization procedure [10, 11, 13, 14] is needed to optimize the sensors performance without affecting the efficiency.

There are various types of sensors that vary slightly in terms of functionality provided by them [15]. A sensor node may use mechanical, biological, chemical, or magnetic sensor to meet the requirements of industry, military applications, precision agriculture applications and healthcare monitoring applications with ease of implementation and maintenance cost. Sensor nodes have two fold capabilities. Primarily they are capable to sense variety of environmental conditions such as pressure, sound level, temperature, humidity variations, vibrations etc. Secondly they also have computing capability. To provide communication between nodes, multi hop communication is used. This facilitates fast transmission of sensed data from source node to respective destination or to specific master node. The network topology and the nodes distributions depend on the nature of the application [16–21]. It is difficult to sustain

M. Elhoseny and A. E. Hassanien, *Dynamic Wireless Sensor Networks*, Studies in Systems, Decision and Control 165, https://doi.org/10.1007/978-3-319-92807-4_5

large operation on sensor nodes for long running time since sensors are operated on battery. It is also very difficult to recharge or change batteries for the sensor nodes. Sensor node poses many constraints like limited amount battery power and difficulty of its recharging if used in harsh environment, limited storage capability, dealing with interoperability problems, etc. Sensor nodes low power, low processing capability, demands energy conservation by avoiding unnecessary sensing activities by development of proper scheduling methods. Another biggest challenge in WSNs is better design of routing protocol because of several network constraints such as random node deployment, energy consumption, data delivery models, node capability, network dynamic and data aggregation.

To enhance the network availability and scalability, several network topologies have been devised. One of them is the clustering model in which the sensors are grouped in clusters. This model improves the performance of the network regarding its lifetime [22]. In a clustering WSN, each cluster usually consists of at least one surrogate node, often referred to as the cluster head (CH). The CH may be dynamically decided or preassigned by the network designer. The communication between the cluster and the base station is facilitated by this CH. The correct choice of CHs will affect directly the network lifetime. In WSNs, the network lifetime is often the most important issue. Besides minimizing the energy consumption of data acquisition by advancing the sensing technology, it is preferable that sensor nodes in a WSN have equal lifetimes so that the entire network achieves a maximum availability. To balance the availability of individual nodes and the entire network, selecting appropriate CHs becomes a key step. There have been many methods proposed to address this need [23–28].

Many studies about energy efficiency of WSN have appeared recently in the literature [24, 26, 29], and several protocols and algorithms have been proposed in order to obtain an optimal power conservation in cluster-based WSN [23, 27]. But most of them do not provide a long lifetime system in a dynamic environment, where the sensors are unattended and it is very difficult to recharge their batteries. Low-Energy Adaptive Clustering Hierarchy (LEACH) is one of the most promising cluster-based routing protocols [30]. LEACH is a distributed single-hope clustering algorithm. It is assumed that all nodes can perform long-distance transmissions to the BS. Thus, selecting the CH depends on the energy used by all clusters. HEED is another cluster-based algorithm [31] based on the nodes' remaining energy and the nodes' neighbors.

The problem of most of the previous methods is the assumption that the number of clusters are fixed during all rounds. This assumption does not guarantee getting the appropriate number of CHs that maximizes the network lifetime. In addition, in many applications we can not specify the number of clusters in advance. Also, in case of a dynamic WSN environment in which the sensor nodes may change their location from time to time, the assumption of fixed structure is not efficient. So, we propose a new scheme for constructing the network structure in a dynamic environment and selecting the CH using Genetic Algorithm [32–36]. In our proposed method, the structure of the network may differ after each round depending of the nodes' characteristics. These characteristics will determine the CHs and the number of the clusters in the network as well. These characteristics are:

- F_1: Average distance.
- F_2: Vulnerability index.
- F_3: Remaining battery power.
- F_4: Number of neighbors.
- F_5: Expected consumed energy.

In addition, there are a set of assumptions that we used in our model such as:

- All nodes are energy-constrained.
- All clusters are homogeneous but they may have different sizes. An homogeneous cluster means all nodes have the same characteristics.
- There is only one BS.

Section 5.2 which describes the proposed model to show how it forms the network clusters. in addition, Sect. 5.3 describes the proposed algorithm and its parameters. Finally, in Sect. 5.4 we discuss the results and the corresponding validations.

5.2 WSN Construction

5.2.1 Related Work

Sensor nodes in the given sensing area are grouped into different clusters, considering their nearest distance to reach others. Here cluster based approach for routing has been used and node with maximum present energy has been selected as cluster head and remaining nodes as cluster members. Thus in each cluster, node with maximum remaining energy is elected as cluster head (CH).

There exist many algorithms that try to maximize the network lifetime [37]. Here we assume that the network is secure and there are no additional security procedures required [38–42]. MODLEACHST [43] is a new method based on LEACH that provides maximum network life time. But their results are constrained to a limited number of transmissions. M-GEAR [44] divides the sensor nodes into four logical regions on the basis of their location in the sensing field. M-GEAR selects cluster heads (CHs) in each independent region. However, in this method the CHs are selected on probabilistically. Also, the Stable Election Protocol (SEP) is a protocol that gives a weighted probability to each node that becomes a CH [45]. In DEEC [46], the existing energy in each node follows a criterion to become a CH. Threshold Sensitive Stable Election Protocol (TSEP) [47] is another method but it is designed for heterogeneous network. The Developed Distributed Energy-Efficient Clustering (DDEEC) [48] permits to balance the cluster head selection over all network nodes following their residual energy.

Intelligent algorithms provide adaptive mechanisms that exhibit intelligent behavior in complex and dynamic environments like WSNs [49]. Various studies [49–53] discuss the routing protocols in cluster-based WSN that rely on intelligent algorithms. In GA-WCA, the load-balanced factor is considered as one of the weights, along with a sum of distances from all neighbor nodes to cluster heads. LA2D-GA takes only the distance as a parameter to calculate the fitness function; however, the representation of a chromosome is a two-dimensional grid which implies valid statistics of a WSN [54].

In most of the studies that use GA to manage WSN, the main objective is to select the CH with the assumption that the number of clusters is determined in advance. Additionally, the process of CH selection depends on only four factors for each node as described in Refs. [50, 51, 53, 55]. These factors are the distance of a node from the cluster centroid, the remaining battery power, the degree of mobility, and the vulnerability index. In other recent studies such as [52], only two parameters have been considered for cluster head selection (number of neighbors and residual energy). But we believe that these factors are not enough to determine the CH selection process using GA. In this contribution we improve the network lifespan by proposing two additional parameters. The additional parameters are (i) the consumed energy of the node and (ii) the number of neighbors' nodes. The first one will indicate the amount of data that the node will be able to transmit and receive in the future, whereas the second one is introduced because receiving and transmitting data consumes energy depending on the data size. also, the CH will manage all data between the BS and all its cluster nodes. The aim of adding these two factors is to increase the network lifetime. This is so because choosing the most appropriate CH will affect the performance of the network. Also, we shall use GA to retrieve the most appropriate structure of the network. This is more effective specially in a dynamic environment in which the number the clusters is not fixed.

GAs are adaptive search methods that simulate some of the natural processes: selection, information, inheritance, random mutation and population dynamics [56]. A GA starts with a population of strings and thereafter generates successive populations of strings. A simple GA consists of three operators [50]: crossover, reproduction and mutation. The chromosome of the GA contains all the building blocks of the solution of the given problem in a form that is suitable for the genetic operators and the fitness function.

In our problem, each proposed network structure is represented as a chromosome. The structure is proposed according to the features of sensor nodes. Our work is built upon the following hypothesis, namely, that the main factors of selecting a node to be CH are the average distance between the node and all its neighbors, the vulnerability index of the node, the degree of mobility, the remaining battery power, the number of neighbors, and the consumed energy. Also, the degree of priority is different for each factor. Therefore, we assign a specific weight for each factor depending on the application according to which the network is designed. Finally, using GA to construct the network structure with these factors will produce the best CHs counts and improve the network lifetime.

5.2.2 Routing Protocols and Cluster Head Selection

Ad hoc routing protocols can be divided into two main categories: topology based and position based. Topology based routing protocols use the information about the links to perform packet forwarding that exists in the network. Position-based routing protocols are used in the geographical position of nodes to take routing decisions that may result in improving efficiency and performance. Topology-based routing can be further divided into two approaches: Proactive and reactive approach. Proactive routing protocols in ad-hoc network periodically broadcast control messages in an attempt to have each node always know a current route to all destinations. But it maintains routing information about the available paths even if these paths are not currently used. Some Proactive routing protocols are Destination Sequenced Distance vector (DSDV), Source-Tree Adaptive Routing (STAR) and Wireless Routing Protocol (WRP). But reactive routing protocols maintain only the routes that are currently in use thereby reducing the burden on the network are appropriate for wireless environments because they initiate a route discovery process when data packets need to be routed. They include the Dynamic Source Routing (DSR) protocol, Ad-hoc On-demand Distance Vector (AODV) protocol, Ad–hoc Ondemand Multi–path Distance Vector (AOMDV) protocol and the Temporally Ordered Routing Algorithm (TORA) protocol.

From the study of related work, one important aspect that has to be considered is, achieving QoS parameters like data accuracy, minimized delay and minimized energy consumption to improve life time of sensor node. Optimization [57–59] is a procedure through which the best possible ways of decision variables are obtained under the given set of constraints and in accordance to a selected optimization objective function. In order to broaden the applicability of the optimization approach to various problem domains, physical and natural principles are used to develop robust optimization algorithms. Simulation Annealing (SA), Evolutionary Algorithms' (EA), Particle Swarm Optimization (PSO) and Ant Colony Optimization (ACO) are few examples of such algorithms. The efficiency of multipath routing depends on the construction and physical distribution of paths. Path selection decides the performance of the multipath routing. If the selected paths are independent of each other or cooperative, multipath routing could improve the performance. Thus, node-disjoint path routing protocols are proposed to avoid interference among paths Multipath routing protocols can provide fault tolerance by having redundant information routed to the destination via alternative paths. Using node-disjoint paths, one can address faulttolerance as well as load sharing. Finding node disjoint paths in case of ad hoc networks is a challenging task. This is due to the fact that the protocol in ad hoc environment should be localized and distributed. If the topology of the network is known, there exist standard methods to discover node-disjoint paths between a given pair of nodes. A protocol may or may not identify all node-disjoint paths between a given pair of nodes.

Given the coordinates of sensor nodes, the distance between two nodes can be approximated by the Euclidean distance (ED). That is, the distance factor can be

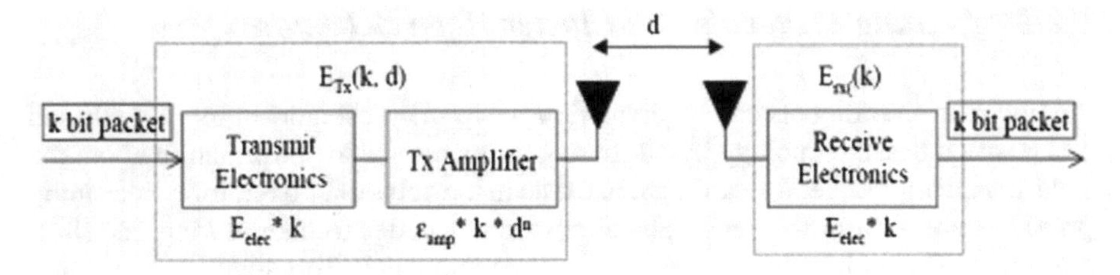

Fig. 5.1 Radio energy dissipation model

calculated as the sum of all distances between the node and its neighbors and the distance between it and the BS. For example, the distance factor D of node v with number of neighbors K can be represented as follows

$$D_v = \sum_{j=1}^{k} ED_{v,j} + ED_{v,BS} \tag{5.1}$$

The consumed energy E, represents the energy required to transfer the aggregated message from the cluster to the BS. According to the first order radio model [53] shown in Fig. 5.1, the consumed energy to send a message with k bits length in d distance is computed via.

For a cluster with k member nodes, cluster transfer energy is defined as [60, 61].

$$E = \sum_{j=1}^{k} E_{T_{jh}} + kE_R + E_{T_{hs}} \tag{5.2}$$

The first part of Eq. 5.2 shows the energy consumed to transmit messages from k member nodes to the cluster head. The second part shows the energy consumed by the cluster head to receive k messages from the member nodes. Finally, the third part represents the energy needed to transmit from the cluster head to the BS. The energy consumed ($E_{T_{ij}}$) needed to transmit a message of length l bits from a node i to a node j and the energy consumed in receiving the l-bit message (E_R) are given, respectively, by

$$E_{T_{ij}} = lE_e + l_{\epsilon_l} d_{ij}^4, \tag{5.3}$$

$$E_{T_{ij}} = lE_e + l_{\epsilon_s} d_{ij}^2, \tag{5.4}$$

and

$$E_R = lE_e + lE_{BF}, \tag{5.5}$$

where d_{ij} is the distance between nodes. However, for a long range transmission such as from a CH to the BS, the energy consumed is proportional to d^4. The EBF represents the cost of the beam forming approach to reduce the energy consumption. Also, the remaining battery power of a node v (denoted by $RP(v)$) can be calculated by using the following expression

$$E_{RP(v)} = E_s - \sum_{t=0}^{c}(E_{Tv}(t) + E_{Rv}(t)), \tag{5.6}$$

where Es is the initial energy of the node, and c the current time. The number of neighbors of a node v (denoted by $N(v)$) is the total number of nodes whose distance d from v is less than or equal to β (β differs from one application to another according to the span of the WSN)

$$N(v) = |n_i| \Leftrightarrow d_{vn_i} \leq \beta : (v \neq n_i \& \beta > 0) \tag{5.7}$$

Finally, based on [62], a node that possesses a high vulnerability values implies that the failure of that node may lead to the disconnection of the entire network. Thus, we should avoid this node to become CHs. The vulnerability index VI of the node v can be calculated as per Eq. 5.8.

$$VI_v = \frac{N^v_{before}}{N^v_{after}} \times \frac{L^v_{before} + 1}{L^v_{after} + 1}, \tag{5.8}$$

where, N^v_{before} is the number of nodes before removing the v node, N^v_{after} is the number of nodes after removing the v node, L^v_{before} is the number of levels before removing the v node, and L^v_{after} is the number of levels after removing the v node.

5.3 The Proposed GA Clustering Method

Each network structure is represented by a set of bits. The number of bits in the structure is equal to the number of network nodes. Each node is represented with either 0 or 1 depending to its role in the network structure. The bit with 1 represents a CH, and the bit with 0, a normal sensor node. So, the number of nones represents the number of clusters. An example of a chromosome for a network with 15 nodes and 5 clusters is shown in Fig. 5.2.

After generating a set of random network structures, GA will calculate the fitness of each structure $f_{\text{chromosome}}$ using Eq. 5.9. Recall that the goal here is to maximize $f_{\text{chromosome}}$

$$f_{chromosome} = [\sum_{i=1}^{|CH|} f_{ch_i}] + \frac{1}{|CH| + |NCH|}, \tag{5.9}$$

Node ID:	1	2	3	4	5	6	7	8	9	10	11	12	13	14	15
Value	1	0	0	1	0	0	0	0	1	1	0	0	1	0	0

Fig. 5.2 Chromosome representation

where N is the number of nodes, $|CH|$ is the number of clusters in the proposed network structure and NCH is the node that has the following two proprieties: firstly, it is selected to be a CH (there are no other nodes linked to it inside the cluster); and secondly, the distance between it and the nearest CH is less than or equal to β.

The value of f_{ch} function is calculated for each cluster. Its values depends on the five features that we propose to select the CH. Let us assume that, each feature i has a specific weight w_i that indicates its priority according to the application. f_{ch} can be cast as given in Eq. 5.10

$$f_{ch} = \frac{1 + \sum_{i=3,4} W_i F_i}{1 + \sum_{i=1,2,5} W_i F_i}. \tag{5.10}$$

In addition to the fitness function, we shall calculate the probability of selecting $p(s_i)$ for each chromosome as given in Eq. 5.11. The value of $p(s_i)$ will determine the chance of the chromosome i to be selected during the crossover or the mutation process. Its formal expression is given by

$$p(s_i) = \frac{f(i)}{\sum_{x=1}^{n} f(x)}, \tag{5.11}$$

where n is the number of the chromosomes in the population. A generation of the GA begins with reproduction. We select the mating pool of the next generation using the weighted roulette wheel.

The algorithm works by generating the initial population of chromosomes of size N according to the network information such as the number of nodes, features weights, and GA parameters. Each chromosome contains a random number of CHs. For each chromosome, we calculate its fitness by computing all node features and assigning each node to a cluster. Then the algorithm calculates the fitness function for each chromosome and applies the roulette wheel to compute the actual number that selects the chromosome. After that, the processes of mutation, crossover and reproduction will be performed according to their ratio. These processes will be repeated until the termination condition is achieved. After that, the best structure of the network will be selected according to the fitness value.

5.4 Cluster Head for Multi-hope Clustering Model

The multi-hope clustering model is a special case of the clustering model in which a CH can not transmit the data directly to the BS for some reasons, i.e. long distance. In this case, the CH tries to fined another CH closer to the BS to works as gateway between them. Now, assume that, we have a WSN contains a set of clusters that are formed by the above CH selection model. Each cluster has a specific CH for a fixed round of data transmission. After each round, the BS selects a CH for each cluster depending on the available information about all nodes such as the remaining energy, location, and neighbors. Also assume that, some of CHs cannot transmit data directly to the BS if the distance between them is more than α. Each of them must transmit its data to an intermediate CH between it and the BS. So, we propose a simple algorithm for constructing the multi-hop model after creating the clusters and determining the CH for each one. The proposed algorithm is based on A-Star (A^*) Algorithm []. The objective of this algorithm is to extend the network lifetime by choosing the path with minimum energy consumption. We assume that, this algorithm will runs at the BS after each round of sending data to it from all CHs to reconstruct the network hierarchy. The BS will run it for CH that far away from it more than a threshold value α. It evaluates a CH n by combining $g(n)$, the cost to reach the CH, and $h(n)$, the cost to get from the CH to the BS. The algorithm selects the CH with minimum $f(n)$. Let the initial CH that we need to determine the best path from it to the BS is i and the remaining energy of CH n is E_n and the count of hindrances between i and n is $|H_i, n|$, the cost function can be calculated as Eqs. (5.12), (5.13) and (5.14).

$$f(n) = g(n) + h(n) \tag{5.12}$$

$$g(n) = ED_{i,n} + E_n + |H_i, n| \tag{5.13}$$

$$h(n) = ED_{n,BS} + |SN_n| \tag{5.14}$$

where $|SN_n|$ refers to the count of nodes in the cluster on n. The number of nodes of the cluster is included into calculation to indicate the traffic loads of the CH. the ED is the energy aware distance that represents the expected distance between two nodes. The ED can be calculated according to Fig. 5.3. After several iterations, the algorithm selects the path $\varrho[x, BS]$ from the CH x to the BS, which is the minimum energy path. The protocol assumes that each CH can reach to any other CH in its environment as shown at Algorithm 3 where N is the count of the CHs.

Figure 5.3 shows an example of the multi-hope intelligent algorithm. The example contains three clusters and one BS. For each cluster, one CH is defined with its features. The features of the CH are (ID, coordinates, Remaining Energy, and Number of Nodes in its cluster) respectively as shown in the figure. Assume that, the CH C cannot interact directly with the BS. It needs another CH to works as a mediator

Algorithm 3: Multi-hope Clustering Algorithm

[t]
1: Initialize α;
2: **for** $x = 0, 1, \ldots, CH_s$ **do**
3:   **if** $ED_{x,BS} \leq \alpha$ **then**
4:     $\varrho[x, BS] = \Phi$, Where Φ represents the direct link between x and BS
5:   **else**
6:     **for** $i = 0, 2, \ldots, N$ **do**
7:       $h(i) = ED_{i,BS} + |SN_i|$
8:       $g(i) = ED_{i,x} + E_i + |H_i, x|$
9:       $f(i) = g(i) + h(i)$
10:     **end for**
11:     $\varrho[x, BS]$ = Min(f.items), where the minimum distance between CH and all other CHs
12:   **end if**
13: **end for**
14: Return the Multi-hope Structure;

between them. Depending to its signal strength, it can select between A and B. Applying our proposed algorithm, C will calculates the cost function $f(A)$ and $f(B)$ for A and B, respectively as shown at Table 5.1.

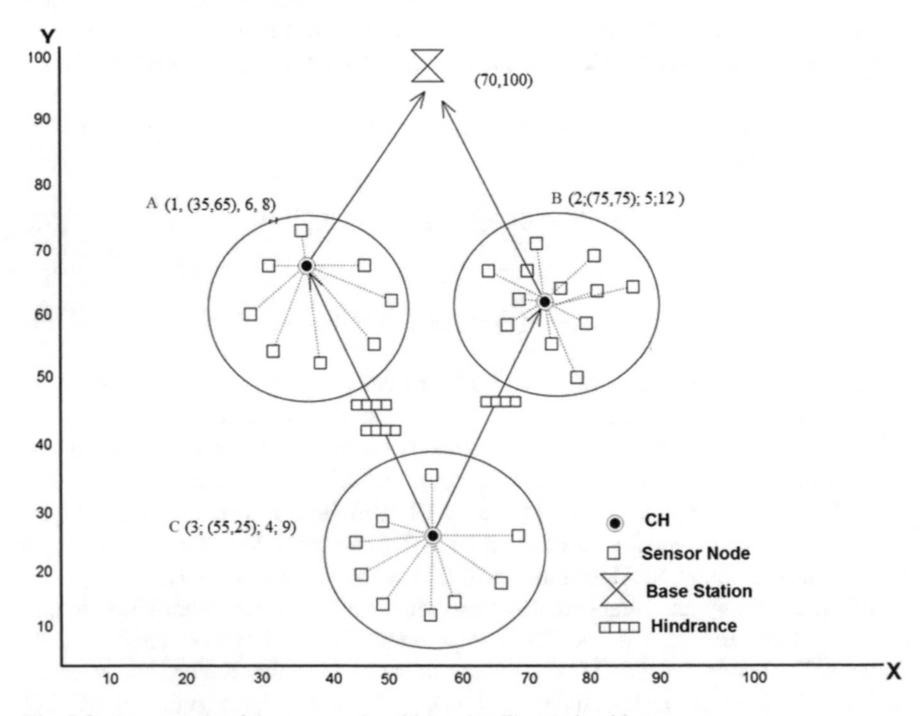

Fig. 5.3 An example of the proposed multi-hop intelligent algorithm

Table 5.1 Cost function for CHs. Where ED_{BS} is the energy aware distance from the CH to the BS while ED_C is the energy aware distance between it and the CH C

CH	E	$\|SN\|$	ED_{BS}	ED_C	$\|H\|$	$g()$	$h()$	$f()$
A	6	8	49.49	44.72	2	62.72	57.49	120.21
B	5	12	25.49	53.85	1	59.85	37.49	97.34

From Table 5.1, the node C will select the CH B to be its parent because it has the minimum cost value.

Given the coordinates of sensor nodes, the energy aware distance ED between two of them can be approximated taking into account their remaining energy. The values of ED is computed by combining the Euclidean distance between the nodes and the environmental hindrances. The hindrances between the node and all other nodes in its cluster such as trees or buildings are taken into account. To calculate the ED between the node V and the node B, we assume that the Euclidean distance between B and V is D_{VB}. Also, assume that the Euclidean distance is used to calculate the distance D_{VC} between V and its CH. Each hindrance between V and any other nodes in the cluster is H and the total number of hindrance inside the cluster is nh. The number of nodes in the cluster is n. Each hindrance has a weight W which represents its ability to consume the node energy. Let $W = \{w_1; w_2; w_3; .w_{nh}\}$ where each H has an ID that is represented by index i of w_i. So, the ED can be represented as Eq. (5.15)

$$ED_V = D_{VC} + \sum_{i=1}^{n} \sum_{j=1, i \neq j}^{n} \sum_{ik=0}^{nh} [H_{ij} * w_i] + D_{VB} \tag{5.15}$$

In case of calculating the distance between a CH i and the BS, ED can be calculated with simple Euclidean distance ($ED_{i,BS} = Di, BS$). Algorithm 4 shows the main steps of the proposed ED algorithm.

Algorithm 4: Energy Aware Distance Algorithm

1: Get the location coordinates for: $B, V, C,$
2: Set $e = 0$;
3: Calculate $DVB = \sqrt{(x_B - x_V)^2 + (y_B - y_V)^2}$
4: Calculate $DVC = \sqrt{(x_C - x_V)^2 + (y_C - y_V)^2}$
5: **for** $i = 1, 2, \ldots, N$ **do**
6:   **for** $j = 1, 2, \ldots, N$ **do**
7:     **for** $k = 1, 2, \ldots, nh$ **do**
8:       $e = e + W[k] * H_{ij}$
9:     **end for**
10:   **end for**
11: **end for**
12: $ED = (e + ED_{V,BS} + ED_{V,C})$
13: return ED

5.5 Experimental Results and Discussion

The proposed algorithm has been programmed and tested using the MS Visual C# 2010 development environment. First, two simulation configurations have been executed 8 times to test the network lifetime and throughput for our method compared with some traditional and recent methods. Each configuration represents different environment characteristics as previously discussed. Finally, a third simulation has been run against some of the GA-based algorithms in WSN. The results show that the network lifetime and throughput is maximized compared with existing protocols. The network parameters for the first simulation are shown in Table 5.1, and the output results in Table 5.2 and Fig. 5.4.

Table 5.2 Network parameters for first simulation

Network parameters		Value
Number of sensors		100
Base station position		(50, 50)
Network area		$100\,\mathrm{m}^2$
Initial energy of each sensor		1.0 J
Idle state energy consumption		50 nJ/bit
Data aggregation energy consumption		5 nJ/bit/report
Amplification energy	$d \geq d_0$	$10\,\mathrm{pJ/bit/m}^2$
Amplification energy	$d < d_0$	$0.0013\,\mathrm{pJ/bit/m}^2$
Amplification energy	$d \geq d_1$	$E_{fs}/10 = E_{fs1}$
Amplification energy	$d < d_1$	$E_{mp}/10 = E_{mp1}$
Packet size		400 bit

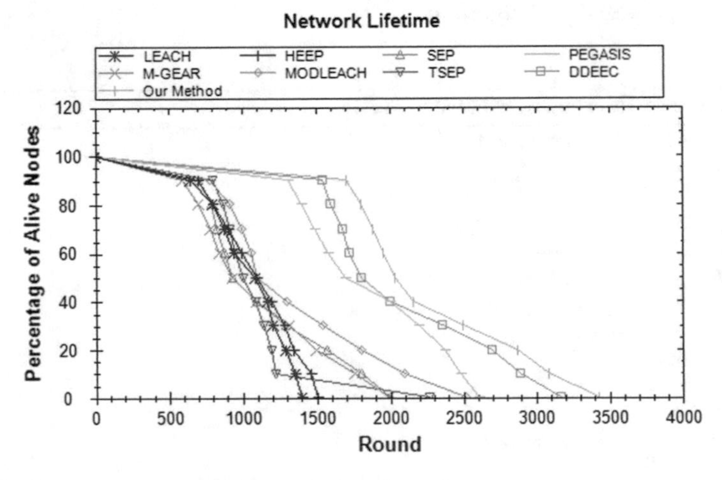

Fig. 5.4 Network lifetime comparison using first simulation parameters

The network parameters for the second simulation are shown in Table 5.3, and the results of its performance are shown in Table 5.4 and Fig. 5.5

The network parameters for the third simulation are the same as in the first simulation. We used the following GA parameters in the third simulation: population size = 20, number of generations = 30, crossover ratio = 0.8, mutation ratio = 0.006, and number of rounds = 1000. The results of the third simulation are shown in Table 5.5.

Using the first simulation parameters also, the results show a good improvement in the network throughput shown in Table 5.6 and Fig. 5.6 (Table 5.7).

Table 5.3 Results of first simulation

Algorithm	Round node die		
	First die	50%	Last die
LEACH	523	1084	1402
HEED	603	1100	1510
SEP	674	940	2003
PEGASIS	1251	1693	2611
M-GEAR	512	932	1989
MODLEACH	668	1101	2517
TSEP	711	1004	2271
DDEEC	1495	1801	3169
OUR METHOD	1591	2037	3424

Table 5.4 Network parameters for second simulation

Network parameters		Value
Number of sensors		400
Base station position		(450, 450)
Network area		400 m^2
Initial energy of each sensor		1.0 J
Idle state energy consumption		50 nJ/bit
Data aggregation energy consumption		5 nJ/bit/report
Amplification energy	$d \geq d_0$	10 pJ/bit/m^2
Amplification energy	$d < d_0$	0.0013 pJ/bit/m^2
Amplification energy	$d \geq d_1$	$E_{fs}/10 = E_{fs1}$
Amplification energy	$d < d_1$	$E_{mp}/10 = E_{mp1}$
Packet size		400 bit

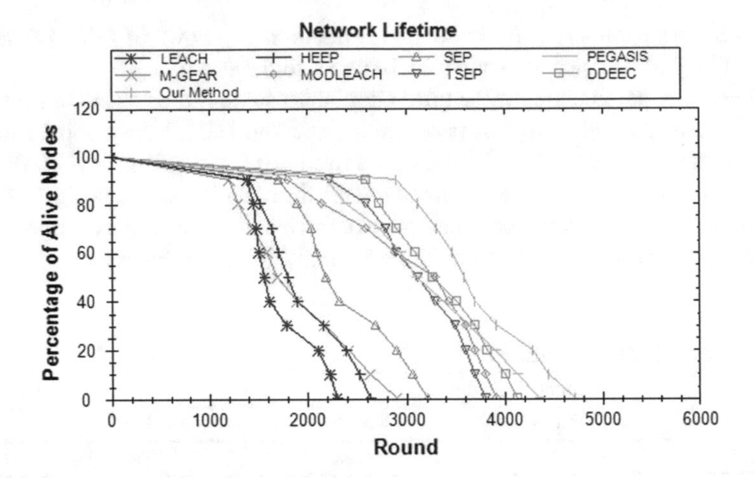

Fig. 5.5 Network lifetime comparison using second simulation parameters

Table 5.5 Results of second simulation

Algorithm	Round node die		
	First die	50%	Last die
LEACH	1232	1570	2300
HEED	1345	1804	2643
SEP	1571	2189	3210
PEGASIS	2065	3100	4356
M-GEAR	1090	1700	2900
MODLEACH	1800	3290	3913
TSEP	2100	3120	3800
DDEEC	2460	3273	4120
OUR METHOD	2601	3586	4712

Table 5.6 Results of third simulation to show the round at which the first node die (FND) and the round time in seconds

Algorithm	FND	50%	Round time	Total time(M:Sec)
HCR	933	1104	2	29:51
OUR METHOD	1591	2037	1.5	24:32

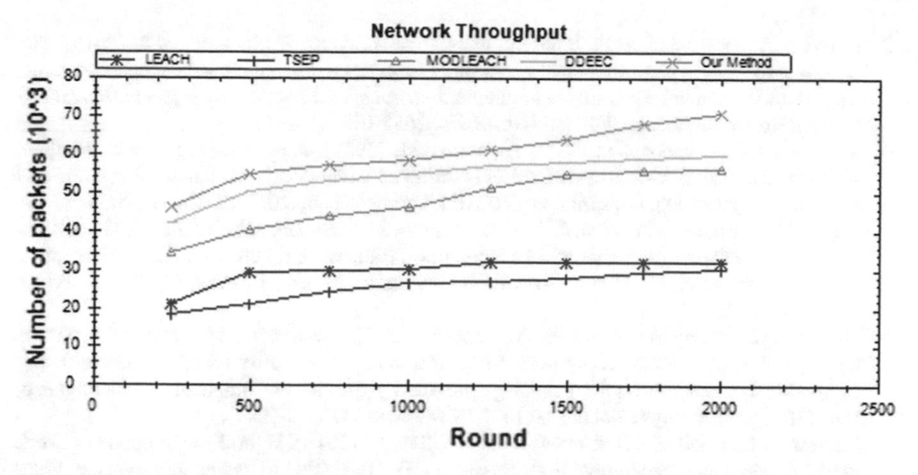

Fig. 5.6 Network throughput comparison using first simulation parameters

Table 5.7 Results of network throughput

Algorithm	Round (500)	Round (1500)	Round (2000)
LEACH	$2.9\,10^4$	$3.2\,10^4$	$3.2\,10^4$
TSEP	$2.1\,10^4$	$2.8\,10^4$	$3\,10^4$
MODLEACH	$4\,10^4$	$5.5\,10^4$	$5.6\,10^4$
DDEEC	$5\,10^4$	$5.8\,10^4$	$6.1\,10^4$
OUR METHOD	$5.5\,10^4$	$6.1\,10^4$	$7.6\,10^4$

5.6 Conclusion

A new method to construct WSN structure that optimize its throughput is proposed. To avoid energy consumption, the proposed algorithm will run after each round at the BS according to the current status of each node within the network. The proposed algorithm has been tested against other existing algorithms using different environment characteristics. The results show that our proposal leads to longer network lifetimes and throughputs, as opposed to other existing methods.

References

1. Elhoseny, M., Abdelaziz, A., Salama, A. S., Riad, A. M., Muhammad, K., & Sangaiah, A. K. (2018). A hybrid model of internet of things and cloud computing to manage big data in health services applications. *Future Generation Computer Systems*. Elsevier. (in Press).
2. Abdelaziz, A., Elhoseny, M., Salama, A. S., & Riad, A. M. (2018). A machine learning model for improving healthcare services on cloud computing environment. *Measurement*, *119*, 117–128. https://doi.org/10.1016/j.measurement.2018.01.022.

3. Darwish, A., Hassanien, A. E., Elhoseny, M., Sangaiah, A. K., & Muhammad, K. (2017). The impact of the hybrid platform of internet of things and cloud computing on healthcare systems: Opportunities, challenges, and open problems. *Journal of Ambient Intelligence and Humanized Computing*. Springer. https://doi.org/10.1007/s12652-017-0659-1.
4. Yuan, X., Li, D., Mohapatra, D., & Elhoseny, M. (2017). Automatic removal of complex shadows from indoor videos using transfer learning and dynamic thresholding. *Computers and Electrical Engineering*. https://doi.org/10.1016/j.compeleceng.2017.12.026. (in Press)
5. Sajjad, M., Nasir, M., Muhammad, K., Khan, S., Jan, Z., Sangaiah, A.K., Elhoseny, M., & Baik, S.W. (2017). Raspberry Pi assisted face recognition framework for enhanced law-enforcement services in smart cities. *Future Generation Computer Systems*. Elsevier. https://doi.org/10.1016/j.future.2017.11.013.
6. Shehab A., Elhoseny M., El Aziz M. A., Hassanien A. E. (2018) Efficient schemes for playout latency reduction in P2P-VoD systems. In: A. Hassanien, & D. Oliva (Eds.), *Advances in soft computing and machine learning in image processing*. Studies in Computational Intelligence, Vol. 730. Springer. https://doi.org/10.1007/978-3-319-63754-9_22.
7. Elhoseny, M., Nabil, A., Hassanien A. E., & Oliva, D. (2018). Hybrid rough neural network model for signature recognition. In A. Hassanien, D. Oliva (Eds.), *Advances in soft computing and machine learning in image processing*. Studies in Computational Intelligence, Vol. 730. Cham: Springer. https://doi.org/10.1007/978-3-319-63754-9_14.
8. Abdeldaim, A. M., Sahlol, A. T., Elhoseny, M., & Hassanien, A. E. (2018). Computer-aided acute lymphoblastic Leukemia diagnosis system based on image analysis. In A. Hassanien, & D. Oliva (Eds.), *Advances in soft computing and machine learning in image processing*. Studies in Computational Intelligence, Vol. 730. Cham: Springer. https://doi.org/10.1007/978-3-319-63754-9.
9. Elhoseny, M., Yuan, X., Yu, Z., Mao, C., El-Minir, H. K., & Riad, A. M. (2015). Balancing energy consumption in heterogeneous wireless sensor networks using genetic algorithm. *IEEE Communications Letters*, *19*(12), 2194–2197.
10. Tharwat, A., Mahdi, H., Elhoseny, M., & Hassanien, A. E. (2018). Recognizing human activity in mobile crowdsensing environment using optimized k-NN algorithm. *Expert Systems With Applications*. https://doi.org/10.1016/j.eswa.2018.04.017. Accessed 12 April 2018.
11. Tharwat, A., Elhoseny, M., Hassanien, A. E., Gabel, T., & Kumar, A. (2018). Intelligent Bezir curve-based path planning model using chaotic particle swarm optimization algorithm. *Cluster Computing*, 1–22. Springer. https://doi.org/10.1007/s10586-018-2360-3.
12. Sarvaghad-Moghaddam, M., Orouji, A. A., Ramezani, Z., Elhoseny, M., & Farouk, A. (2018). Modelling the Spice parameters of SOI MOSFET using a combinational algorithm. *Cluster Computing*. Springer. https://doi.org/10.1007/s10586-018-2289-6. (in Press).
13. Rizk-Allah, R. M., Hassanien, A. E., & Elhoseny, M. (2018). A multi-objective transportation model under neutrosophic environment. *Computers and Electrical Engineering*. Elsevier. https://doi.org/10.1016/j.compeleceng.2018.02.024. (in Press).
14. Batle, J., Naseri, M., Ghoranneviss, M., Farouk, A., Alkhambashi, M., & Elhoseny, M. (2017). Shareability of correlations in multiqubit states: Optimization of nonlocal monogamy inequalities. *Physical Review A*, *95*(3), 032123. https://doi.org/10.1103/PhysRevA.95.032123.
15. Elhoseny, M., Hosny, A., Hassanien, A. E., Muhammad, K., & Sangaiah, A. K. (2017). Secure automated forensic investigation for sustainable critical infrastructures compliant with green computing requirements. *IEEE Transactions on Sustainable Computing*, *PP*(99). https://doi.org/10.1109/TSUSC.2017.2782737.
16. Elhoseny, H., Elhoseny, M., Riad, A. M., & Hassanien, A. E. (2018). A framework for big data analysis in smart cities. In A. Hassanien, M. Tolba, M. Elhoseny, & M. Mostafa (Eds.), *AMLTA 2018 the international conference on advanced machine learning technologies and applications (AMLTA2018)*, Advances in Intelligent Systems and Computing, Vol. 723. Cham: Springer. https://doi.org/10.1007/978-3-319-74690-6_40.
17. Elhoseny, M., Shehab, A., & Osman, L. (2018). An empirical analysis of user behavior for P2P IPTV workloads. In A. Hassanien, M. Tolba, M. Elhoseny, & M. Mostafa (Eds.), *AMLTA 2018 the international conference on advanced machine learning technologies and applications*

(AMLTA2018), Advances in Intelligent Systems and Computing, Vol. 723. Cham: Springer. https://doi.org/10.1007/978-3-319-74690-6_25.
18. Wang, M. M., Qu, Z. G., & Elhoseny, M. (2017). Quantum secret sharing in noisy environment. In X. Sun, H. C. Chao, X. You, & E. Bertino (Eds.), *Cloud computing and security, ICCCS 2017*. Lecture Notes in Computer Science, Vol. 10603. Cham: Springer.https://doi.org/10.1007/978-3-319-68542-7_9.
19. Elsayed, W., Elhoseny, M., Riad, A. M., & Hassanien, A. E. (2018). Autonomic self-healing approach to eliminate hardware faults in wireless sensor networks. In A. Hassanien, K. Shaalan, T. Gaber, & M. Tolba (Eds.), *Proceedings of the international conference on advanced intelligent systems and informatics 2017*, AISI 2017. Advances in Intelligent Systems and Computing, Vol. 639. Cham: Springer. https://doi.org/10.1007/978-3-319-64861-3_14.
20. Abdelaziz, A., Elhoseny, M., Salama, A. S., Riad, A. M., Hassanien, A. E. (2018). Intelligent algorithms for optimal selection of virtual machine in cloud environment, towards enhance healthcare services. In A. Hassanien, K. Shaalan, T. Gaber, M. Tolba (Eds.), *Proceedings of the international conference on advanced intelligent systems and informatics 2017*, AISI 2017, Advances in Intelligent Systems and Computing, Vol. 639. Cham: Springer. https://doi.org/10.1007/978-3-319-64861-3_27.
21. Shehab, A., Ismail, A., Osman, L., Elhoseny, M., & El-Henawy, I. M. (2018). Quantified self using IoT wearable devices. In A. Hassanien, K. Shaalan, T. Gaber, M. Tolba (Eds.), *Proceedings of the International Conference on Advanced Intelligent Systems and Informatics 2017*, AISI 2017. Advances in Intelligent Systems and Computing, Vol. 639. Cham: Springer. https://doi.org/10.1007/978-3-319-64861-3_77.
22. Elhoseny, M., Elminir, H., Riad, A., & Yuan, X. (2016). A secure data routing schema for WSN using elliptic curve cryptography and homomorphic encryption. *Journal of King Saud University–Computer and Information Sciences*, *28*(3), 262–275.
23. Tyagia, S., & Kumarb, N. (2013). A systematic review on clustering and routing techniques based upon leach protocol for wireless sensor networks. *Journal of Network and Computer Applications*, *36*(2), 623–645.
24. Ali, J., Kumar, G., & Rai, M. K. (2013). Major energy efficient routing schemes in wireless sensor networks. *International Journal of Computers and Technology*, *4*(2), 261–266.
25. Elhoseny, M., Yuan, X., El-Minir, H. K., & Riad, A. M. (2014). Extending self-organizing network availability using genetic algorithm. In *Fifth international conference on computing, communications and networking technologies (ICCCNT)*, (pp. 1–6).
26. Pantazis, N. A., Nikolidakis, S. A., & Vergados, D. D. (2013). Energy-efficient routing protocols in wireless sensor networks: A survey. *Communications Surveys and Tutorials*, *15*(2), 551–591.
27. Du, T., Qu, S., Liu, F., & Wang, Q. (2015). An energy efficiency semi-static routing algorithm for WSNs based on HAC clustering method. *Information Fusion*.
28. Riad, A. M., El-Minir, H. K., & Elhoseny, M. (2013). Secure routing in wireless sensor networks a state of the art. *International Journal of Computer Applications*, *67*(7), 7–12.
29. Yuan, X., Elhoseny, M., El-Minir, H. K., & Riad, A. M. (2017). A genetic algorithm-based dynamic clustering method towards improved WSN longevity. *Journal of Network and Systems Management*, *25*(1), 21–46.
30. Kang, S. H., & Nguyen, T. (2012). Distance based thresholds for cluster head selection in wireless sensor networks. *IEEE Communications Letters*, *16*(9), 1396–1399.
31. Younis, O., & Fahmy, S. (2004). Heed: A hybrid, energy-efficient, distributed clustering approach for ad hoc sensor networks. *IEEE Transactions on Mobile Computing*, *3*(4), 366–379.
32. Hosseinabadi, A. A. R., Vahidi, J., Saemi, B., Sangaiah, A. K., & Elhoseny, M. (2018). Extended genetic algorithm for solving open-shop scheduling problem. *Soft Computing*. https://doi.org/10.1007/s00500-018-3177-y.
33. Metawa, N., Elhoseny, M., Hassan, M. K., & Hassanien, A. E. (2016). Loan portfolio optimization using genetic algorithm: A case of credit constraints. In *2016 12th international computer engineering conference (ICENCO)*, pp. 59–64.
34. Elhoseny, M., Tharwat, A., & Hassanien, A. E. (2017c). Bezier curve based path planning in a dynamic field using modified genetic algorithm. *Journal of Computational Science*. https://doi.org/10.1016/j.jocs.2017.08.004.

35. Metawa, N., Hassan, M. K., & Elhoseny, M. (2017). Genetic algorithm based model for optimizing bank lending decisions. *Expert Systems with Applications*, *80*, 7582. https://doi.org/10.1016/j.eswa.2017.03.021.
36. Elhoseny, M., Shehab, A., & Yuan, X. (2017). Optimizing robot path in dynamic environments using genetic algorithm and Bezier curve. *Journal of Intelligent & Fuzzy Systems*, *33*(4), 2305–2316. IOS-Press. https://doi.org/10.3233/JIFS-17348.
37. Elhoseny, M., El-Minir, H. K., Riad, A. M., & Yuan, X. (2014). Recent advances of secure clustering protocols in wireless sensor networks. *International Journal of Computer Networks and Communications Security*, *2*(11), 400–413.
38. Elhoseny, M., Ramírez-González, G., Abu-Elnasr, O. M., Shawkat, S. A., Arunkumar, N., & Farouk, A. (2018). Secure medical data transmission model for IoT-based healthcare systems. *IEEE Access*, *PP*(99). https://doi.org/10.1109/ACCESS.2018.2817615.
39. Shehab, A., Elhoseny, M., Muhammad, K., Sangaiah, A. K., Yang, P., Huang, H., & Hou, G. (2018). Secure and robust fragile watermarking scheme for medical images. *IEEE Access*, *6*(1), 10269–10278. https://doi.org/10.1109/ACCESS.2018.2799240.
40. Farouk, A., Batle, J., Elhoseny, M., Naseri, M., Lone, M., Fedorov, A., Alkhambashi, M., Ahmed, S.H., Abdel-Aty, M., (2018). Robust general N user authentication scheme in a centralized quantum communication network via generalized GHZ states. *Frontiers of Physics*, *13*, 130306. Springer. https://doi.org/10.1007/s11467-017-0717-3.
41. Elhoseny, M., Elkhateb, A., Sahlol, A., Hassanien, A. E. (2018). Multimodal biometric personal identification and verification. In A. Hassanien, & D. Oliva (Eds.), *Advances in soft computing and machine learning in image processing*. Studies in Computational Intelligence, Vol. 730. Cham: Springer. https://doi.org/10.1007/978-3-319-63754-9_12.
42. Elhoseny, M., Essa, E., Elkhateb, A., Hassanien, A. E., & Hamad, A. (2018). Cascade multimodal biometric system using fingerprint and Iris patterns. In A. Hassanien, K. Shaalan, T. Gaber, & M. Tolba (Eds.), *Proceedings of the International Conference on Advanced Intelligent Systems and Informatics 2017*, AISI 2017, Advances in Intelligent Systems and Computing, Vol. 639. Cham: Springer. https://doi.org/10.1007/978-3-319-64861-3_55.
43. Mahmood, D., Javaid, N., Mahmood, S., Qureshi, S., Memon, A. M., & Zaman, T. (2013). Modleach a variant of leach for WSNS. In *Eighth international conference on broadband and wireless computing and communication and applications*, (pp. 158–163).
44. Nadeem, Q., Rasheed, M. B., Javaid, N., Khan, Z. A., Maqsood, Y., & Din, A. (2013). M-gear gateway-based energy-aware multi-hop routing protocol for WSNS. In *Eighth international conference on broadband and wireless computing and communication and applications*, (pp. 164–169).
45. Li, Q., & Qingxin, Z. (2006). Design of a distributed energy-efficient clustering algorithm for heterogeneous wireless sensor networks. *Computer Communications*, *29*(12), 2230–2237.
46. Lindsey, S., & Raghavendra, C. S. (2002). PEGASIS: power-efficient gathering in sensor information systems. In *Aerospace conference proceedings*, (Vol. 3, pp. 1125–1130).
47. Kashaf, A., Javaid, N., Khan, Z. A., & Khan, I. A. (2012). TSEP: Threshold-sensitive stable election protocol for WSNS. In *Conference on frontiers of information technology*, (pp. 164–168).
48. Elbhiri, B., Saadane, R., & Aboutajdine, D. (2010). Developed distributed energy-efficient clustering (DDEEC) for heterogeneous wireless sensor. In *Communications and mobile network (ISVC)*, (pp. 1–4), Rabat.
49. Guo, W., & Zhang, W. (2014). A survey on intelligent routing protocols in wireless sensor networks. *Journal of Network and Computer Applications*, *38*, 185–201.
50. Ahmed, G., Khan, N. M., & Ramer, R. (2008). Cluster head selection using evolutionary computing in wireless sensor networks. In *Progress in electromagnetics research symposium*, (pp. 883–886).
51. Bhaskar, N., Subhabrata, B., & Soumen, P. (2010). Genetic algorithm based optimization of clustering in ad-hoc networks. *International Journal of Computer Science and Information Security*, *7*(1), 165–169.

52. Asim, M., & Mathur, V. (2013). Genetic algorithm based dynamic approach for routing protocols in mobile ad hoc networks. *Journal of Academia and Industrial Research*, *2*(7), 437–441.
53. Karimi, A., Abedini, S. M., Zarafshan, F., & Al-Haddad, S. A. R. (2013). Cluster head selection using fuzzy logic and chaotic based genetic algorithm in wireless sensor network. *Journal of Basic and Applied Scientific Research*, *3*(4), 694–703.
54. Rana, K., & Zaveri, M. (2013). Synthesized cluster head selection and routing for two tier wireless sensor network. *Journal of Computer Networks and Communications*, *13*(3).
55. Hussain, S., Matin, A. W., & Islam, O. (2007). Genetic algorithm for energy efficient clusters in wireless sensor networks. In *International conference on information technology*.
56. Sivagami, A., M. Rathnakumar. (2013). Economic generation scheduling using genetic algorithm. *Social Science Research Network*.
57. El Aziz, M. A., Hemdan, A. M., Ewees, A. A., Elhoseny, M., Shehab, A., Hassanien, A. E., & Xiong, S. (2017). Prediction of biochar yield using adaptive neuro-fuzzy inference system with particle swarm optimization. In *2017 IEEE PES PowerAfrica Conference*, (pp. 115–120), June 27–30, 2017. Accra-Ghana: IEEE. https://doi.org/10.1109/PowerAfrica.2017.7991209.
58. Ewees, A. A., El Aziz, M. A., & Elhoseny, M. (2017). Social-spider optimization algorithm for improving ANFIS to predict biochar yield. In *8th international conference on computing, communication and networking technologies (8ICCCNT)*, July 3–5. Delhi-India: IEEE.
59. Metawa, N., Elhoseny, M., Hassan, M. K., & Hassanien, A. E. (2016) Loan portfolio optimization using Genetic Algorithm: A case of credit constraints. In *Proceedings of 12th international computer engineering conference (ICENCO)*, (pp. 59–64). IEEE. https://doi.org/10.1109/ICENCO.2016.7856446.
60. Elhoseny, M., Yuan, X., El-Minir, H. K., & Riad, A. M. (2016). An energy efficient encryption method for secure dynamic WSN. *Security and Communication Networks*, *9*(13), 2024–2031.
61. Elhoseny, M., Elleithy, K., Elminir, H., Yuan, X., & Riad, A. (2015). Dynamic clustering of heterogeneous wireless sensor networks using a genetic algorithm towards balancing energy exhaustion. *International Journal of Scientific and Engineering Research*, *6*(8), 1243–1252.
62. Ahmed, G., Khan, N. M., Khalid, Z., & Ramer, R. (2008). Cluster head selection using decision trees for wireless sensor networks. In *Sensor networks and information processing*, (pp. 173–178).

Part II
WSN for Secure Data Processing and Live Data Aggregation

The second part of this book discusses an important challenge of WSN applications which is secure data processing and live data aggregation. This part aims to show how to protect WSNs from different types of attack. In addition, it provides specific guidelines toward building a secure WSN for dynamic and complex applications. Collecting the live data from a running system without affecting its performance is considered a nontrivial task especially in some modern and distributed systems such as SCADA systems. This part shows how to aggregate such live data using WSNs. Moreover, it reviews and discusses the most prominent secure clustering routing algorithms that have been developed for WSNs. The goal of this review is to provide a detailed discussion for securing techniques that can be applied to secure data transmission in WSN. The requirements toward building a secure clustering model for WSN are extensively explained. This part represents a tool for building a secure data transmission model to deal with limited sensor resources, i.e., memory size and processing power. It shows how the encryption methods can be used for both text and image data encryptions in WSNs environments that need to capture text data as well as images. A data aggregation method that enables the cluster head to aggregate the encrypted data without decrypting first and re-encrypting to avoid energy consumption is also discussed.

Chapter 6
Secure Data Transmission in WSN: An Overview

Abstract Building a secure routing protocol in WSN is not trivial process. Thee are two main types of security attacks against WSNs: active and passive. WSN as a new category of computer-based computing platforms and network structures is showing new applications in different areas such as environmental monitoring, health care and military applications. Although there are a lot of secure data transmission schemes designed for data aggregation and transmission over a network, the limited resources and the complex environment make it invisible to be used with WSNs. Furthermore, secure data transmission is a big challenging issue in WSNs especially for the application that uses image as its main data such as military applications. This problem is mainly related to the limited resources and data processing capabilities. This chapter introduces a secure data processing and transmission schema in WSN. The chapter reviewed and critically discussed the most prominent secure clustering routing algorithms that have been developed for WSNs. Then, we explained the guidelines and the steps towards building a simple solution for securing the dynamic cluster network while consuming as little energy as possible and is adapted to a low computing power. Moreover, four phased towards building a secure clustering algorithm for WSN are proposed. These phases are secure cluster head selection, secure cluster formation, secure data aggregation by the cluster head from its cluster nodes, and secure data routing to the base station. Also, the chapter proposes and applies an evaluation criteria for the existing secure clustering algorithms.

6.1 Overview

Building a secure routing protocol in WSN is not trivial process [11–15]. It looks like an optimization process [16, 17, 19, 20] through which we try to find the optimum solution that maximize the network performance in an environment with a set of complicated constraints. The main purpose is not only to design new routing protocol [21, 22] that guarantee the network efficiency, but also balancing between this efficiency and the security requirements. For that purpose, we designed and followed a general framework to simplify the process of building such that protocol. An

M. Elhoseny and A. E. Hassanien, *Dynamic Wireless Sensor Networks*, Studies in Systems, Decision and Control 165, https://doi.org/10.1007/978-3-319-92807-4_6

overview of the working steps towards building the proposed protocol is described. Then the protocol objectives and methodology are discussed.

Generally, our protocol was proposed as a Genetic Algorithm-based, clustering method that optimizes dynamic node clustering to extend the network lifetime. In GASONeC, the remaining energy, the expected energy expenditure, the distance to the base station, and the number of nodes in the vicinity are employed to search for an optimal, dynamic network structure. Balancing these factors is the key of organizing nodes into appropriate clusters and designating a surrogate node as cluster head. The factors are encoded into the fitness function of Genetic Algorithm (GA). In the optimization process [23–25], each GA chromosome represents a designation map of cluster heads. A gene in a chromosome specifies if the corresponding node serves as a cluster head. Given a cluster head, the node clusters are then formed following the nearest neighbor rule, and the fitness of a WSN structure prescribed by a chromosome is hence determined by the evaluation of all clusters. In each transmission round, network structure is updated dynamically to achieve network longevity. To ensure security in such a dynamically clustered sensor network, the security protocol must also take energy consumption into consideration.

6.2 The Working Steps for Building GASONeC Protocol

Figure 6.1 shows the phases of building a secure routing protocol in details. It contains the main phases that was followed to create our secure routing protocol for WSN. This figure organizes the main steps and their relationship. There are three main phases: Design phase, testing phase, and performance measure phase. In addition, and intermediate layer called 'Security Procedure Layer (SPL)' is proposed to work as a reference for all other phases as described later. Each of these steps contains a set of tasks and is directly related to the remaining phases. Each phase is explained in details as the following.

6.2.1 Designing Phase

The design phase aims to set the main guidelines towards the actual implementation of GASONeC. These guidelines can be determined by following two main tasks. First, determine the main requirements towards building a secure routing algorithm. In order to do that, this task contains three sub-tasks:

- **Choose Network Architecture**: during this task, we explored the existing network structures, such as flat-routing and cluster-routing [26–29], and determined the strengthens and limitations of each. A comparison was conducted between all of them and based on it we decided to work with clustering model.

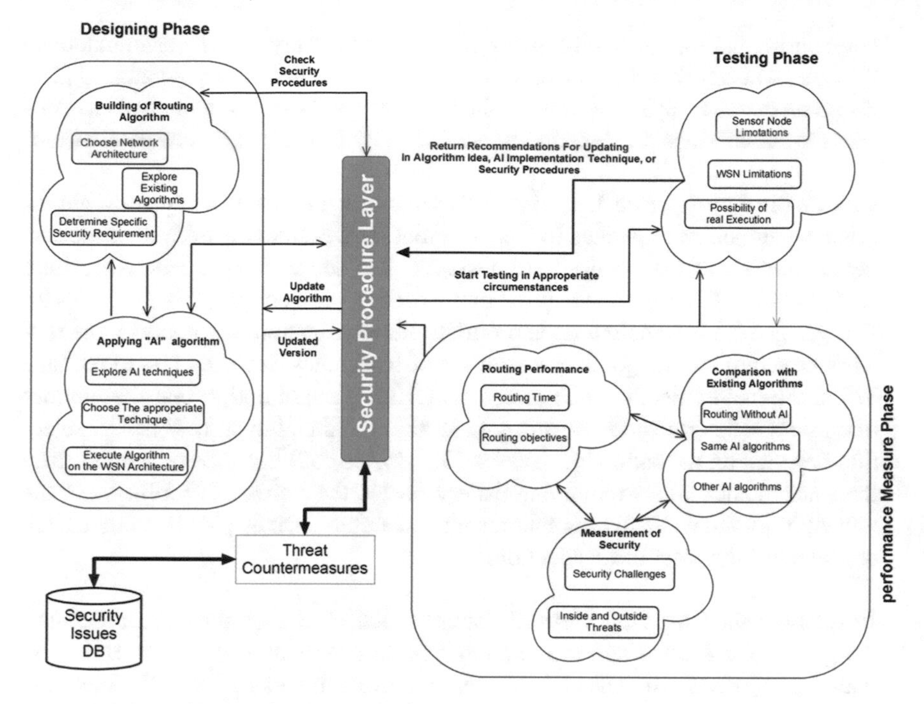

Fig. 6.1 Phases of building a secure routing protocol

- **Explore Existing Routing Algorithms**: the goal of this subtask is to literature the existing algorithms to define their limitations. In addition, a proposed solution for each limitation was provided.

- **Determine Specific Security Requirement**: the security requirements are determined based on the traditional and advanced limitations of WSN. A set of attacks and threats were defined in this subtask. The aim of this subtask is to clarify the threats and limitations towards building a secure routing protocol in WSN.

The second task in the design phase is responsible of choosing the appropriate intelligent technique of building the algorithm. This tasks includes three main subtasks as the following:

- **Explore Intelligent Techniques**: There are a lot of intelligent techniques that are proposed to constructing and organizing WSN's operations [30–32]. Various researches [32–36] discussed the routing protocols in Cluster-base WSN based on intelligent algorithms as reinforcement learning (RL), ant colony optimization (ACO), fuzzy logic (FL), genetic algorithm (GA) [37], and neural networks (NNs). Moreover, Many clustering methods have been proposed. Local Negotiated Clustering Algorithm (LNCA) employs the similarity of data acquired by nodes as a clustering criterion. ACE uses node degree as a feature to group sensor nodes into clusters. GA-WCA uses a load balanced factor and total distance from the neighboring nodes. LA2D-GA takes distance as the only parameter to calculate the fitness; however, a chromosome is represented as a two-dimensional grid [38]. Based on that, a comparison was conducted between the existing methods to evaluate the network performance in terms of security [39–43] and network lifetime.

- **Choose the Appropriate Technique**: This sub-task aims to select the most appropriate technique that matches the requirements and limitations of WSN based on the conclusion of the previous sub-task. We decided to use Genetic Algorithm (GA) [44–46]. GA has been applied in the routing protocol of WSN [35, 47–49]. A key objective is to define an appropriate fitness function that encodes the network structure and its goodness. In most of researches that uses GA to manage WSN, the main objective is to select the CHs with assumption that the number of clusters is determined in advance. In addition to that, the process of CH selection depends on the following factors: [33, 35, 36, 50] distance of a node from the cluster centroid, the remaining battery power, the degree of mobility, and the vulnerability index. While in other recent researches such as [34] less criteria are considered for cluster head selection.
- **Implement the Proposed Method Using the Selected Technique**: At this point, the problem is defined and the implementation technique is clarified. Based on that, the proposed protocol is developed and evaluated using the SPL. First, the dynamic clustering parameters are proposed. Then, GA is used to construct the network as will be discussed later. After that, the cryptography schema is applied to protect the data routing process. The SPL is used to measure if the implementation process matches the security goals.

6.2.2 *Testing Phase*

After finishing the implementation of the proposed protocol, the testing phase starts. It aims to evaluate the implementation details of the proposed algorithm to explore the implementation errors. Based on the individual testing of each task at the design phase, this phase executes complete test to evaluate the algorithm work using the WSN limitations. In addition, it measures to what extend the proposed algorithm can be applied in the real world. Similar to the individual test of each sub-task in the design phase, the testing process consults the SPL to make sure that the algorithm provide the acceptable limit of protection against the pre-defined attacks.

6.2.3 *Performance Measure Phase*

Secure data transmission and extending WSN longevity are our two main goals. In order to build a secure routing protocol that extend the network lifetime, a set of criteria and conditions have to be achieved. The goal of this phase is to evaluate and compare the performance of the proposed protocol with the existing state-of-the art methods using the same parameters and experimental environment. In addition, the protocol efficiency is measured. This phase contains three main tasks:

- **Comparison with Existing Protocols**: This task aims to conduct a comparison with the existing algorithms. The comparison includes both traditional and intelligent algorithms. The evaluation process include a set of points such as the energy consumption, memory requirement, and network lifetime.

- **Measurement of Security Goals**: The goal of this subtask is to test the performance by simulating the work of each attack. This sub-task aims to find the weakness of our protocol to be taken into consideration. Moreover, it tries to evaluate the performance of the network in different environments using different types of threats.

- **Routing Performance**: As we mentioned before, the efficiency is an important evaluation criteria for any protocol. This sub-task evaluates the efficiency of our protocol using the round time. In addition, it tries to find the change of the efficiency in different environments using different data, such as text and images.

6.2.4 Security Procedure Layer

During each task, SPL represents the guide to determine whether the proposed idea follows the security requirements or not. This process is done using a set of predefined attacks and their countermeasures stored at the security database. During the design phase, SPL is used to check and update the work while in testing and performance measure phases it aims to evaluate the performance of the protocol against a set of attacks.

6.3 Security Problems of Wireless Sensor Network

Security attacks against WSNs are categorized into two main branches: active and passive. In passive attacks, attackers are typically hidden and aim to monitor the communication link to collect data. The common examples of passive attacks are eavesdropping, node malfunctioning, node destruction and traffic analysis types. In active attacks, an adversary actually affects the operations in the attacked network. This effect may be the objective of the attack and can be detected. For example, the networking services may be degraded or terminated as a result of these attacks. The common examples of active attacks are Denial-of-Service (DoS), hole attacks, flooding and Sybil types [51]. The source of the attack comes to the network from inside, outside, or both [52]. We can summarize the common types of attacks in WSN as the following:

- **Denial of Service**: It sends unnecessary packets and utilizes more network bandwidth to prevents the user from accessing the service or resource.
- **Selective Forwarding**: It tries to put a malicious node to act as normal node and drop the messages as soon as they receive it.
- **Sinkhole**: This attack adds a node to the network to capture all data as if it was the base station.
- **Sybil**: The malicious node claims multiple identities to be able communicate with many nodes.
- **Wormhole**: This attack records the messages to another location and may retransmit them or a selective part of them.
- **HELLO Flood**: This attack sends the HELLO packet to the nodes, the node may assumes the attacked device as a neighbor that tries to connect with it. It aims to consume the network resources.
- **Spoofed, Altered or Replayed Routing Information**: This is the most direct attack. By spoofing, altering or replaying routing information the attacker can complicate the network through some actions like create routing loops or generating false error messages.
- **Black-Hole**: The malicious node communicates the destination node with false route information to enforce it to send the reply to the malicious node.
- **Node Destruction**: This attack aims either to make the node unavailable to replace it with a malicious one with the same identifier, or to prevent it from collecting data.
- **Monitor and Eavesdropping**: This attack aims to gather information about the network.
- **Traffic Analysis**: This attack aims to intercept and examine messages in order to deduce information from patterns in communication. Its danger comes from its ability to work even when the messages are encrypted.
- **Node Replication**: This attack creates duplicate nodes and built up various attacks using them nodes.

- **Message Corruption**: This attack performs three main actions: receives message, modifies it to be not understandable, and then forwarding it to its destination.

- **Jamming**: Jamming interferes with the radio frequencies of the sensor nodes to make them unavailable.

- **Node Malfunction**: This attack generates inaccurate part of data that could expose the integrity of the data-aggregating process at the CH.

Table 6.1 lists these types of attack with an assigned code for each of them to use it in the evaluation process of the existing methods. In addition, the source and the type of each attack are listed. Based on that, Fig. 6.2 summarize the security challenges of WSN.

These attacks aim to affect the transmitted data with one of the following threats [53]:

- Interruption: is an attack on the availability of the network. Its main aim is to make an asset of the system,i.e., sensor node, unavailable or unusable. Denial of Service attacks [52] have become very well-known example of interruption.
- Interception: is an attack on confidentiality. The sensor network can be compromised by the attacker to gain unauthorized access to sensor node or data store within it. Spoofing attack is a well-known example.

Table 6.1 The common types of attacks of WSN and the source of each

Code	Name	Active	Passive	Inside	Outside
A_1	Denial of service	✓		✓	✓
A_2	Selective forwarding		✓	✓	
A_3	Sinkhole	✓		✓	
A_4	Sybil	✓		✓	
A_5	Wormhole	✓		✓	✓
A_6	HELLO flood		✓	✓	
A_7	Spoofed, altered or replayed routing information	✓		✓	✓
A_8	Black-Hole	✓		✓	
A_9	Node destruction	✓			✓
A_{10}	Monitor and eavesdropping		✓		✓
A_{11}	Traffic analysis	✓			✓
A_{12}	Node replication	✓			✓
A_{13}	Message corruption	✓		✓	✓
A_{14}	Jamming	✓			✓
A_{15}	Node malfunction	✓		✓	

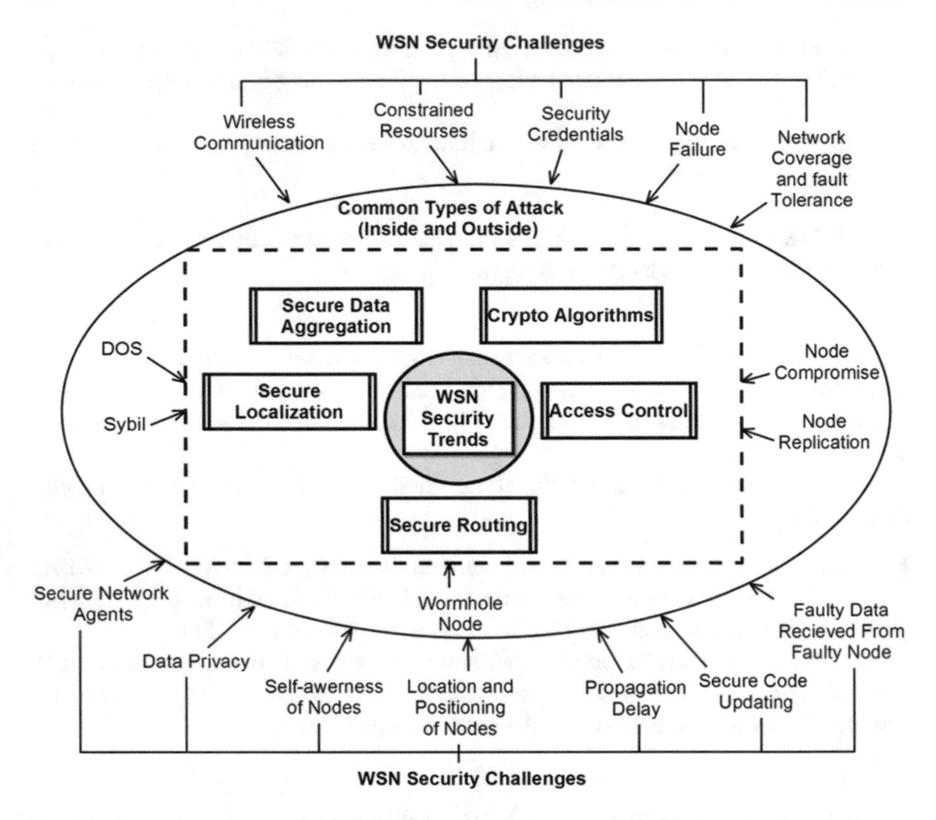

Fig. 6.2 Security challenges for WSN and the common types of attack

- Modification: is an attack on integrity of the system. It this attack unauthorized party not only accesses the data but also modifies the content of a message being transmitted in a network.
- Fabrication: is an attack on authentication in which the attacker make an insertion of messages in a network and tries to make it as it is sent from authorized node.

Methods to address WSN security attacks aim at the following aspects [54]:

- Preventing Attacks: It aims to prevent any attack before it happens. Any proposed technique will have to defend against the targeted attack.
- Detecting Attacks: If an attacker manages to pass the measures taken by the prevention mechanism, the security solution would immediately switch into the detection phase of the counter attack in progress and specifically identify the nodes that are being compromised.
- Removing Attacks: It aims to mitigate any attack after it happens by removing the affected nodes and securing the network.

6.4 Secure Clustering Evaluation Criteria

In this section we discuss the criteria which we will use to evaluate the existing secure clustering method.

6.4.1 Completeness

Secure clustering is a sequential process that must guarantee the security goals, i.e. confidentiality, integrity, and availability; in each phases. This process consists of two main stages: cluster building and data transmission. With more details, the cluster building stage starts with cluster formation in which the CHs are determined and then each node is assigned to CH. The next stage, i.e., data transmission, aims to protect the collected data during its transferring from nodes to the base station. It can be partitioned two main phases: data aggregation, and data routing to base-station. Data aggregation is the process of transmitting data from nodes to the CH inside the cluster. Then CHs forward the data to the base station through a specific path which is known as routing process. Finally, the base station receives the data and extracts the meaning, and then the process will start again as shown at Fig. 6.3. In order to create a secure clustering method, all of these phases must be applied together. Accordingly, we will use these phases to evaluate the existing secure clustering method and show to what extend each of them is a complete secure clustering algorithm or not. We will use $S - CH$, $S - CF$, $S - DA$, and $S - DR$ to indicate to the four phases respectively.

6.4.2 Achieving Security Goals

Secure clustering algorithm must achieve the security goals, i.e., integrity, confidentiality, availability, and freshness to avoid attacks and threats as much as possible. These goals can be summarized as the following [55].

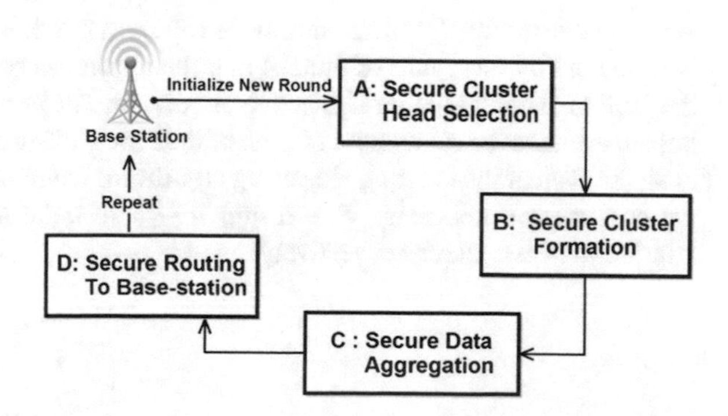

Fig. 6.3 Secure clustering process consists of four steps: Secure Cluster Heads Selection, Secure Clusters Formation, Secure Data Aggregation, and Secure Routing of Data to the Base-station. The arrows depict the data flow

6.4.2.1 Integrity

Data must not be changed in transit, and steps must be taken to ensure that data cannot be altered by unauthorized party. To insure that data reaches to the intended receiver without any alteration, a technique like hash function can be used.

6.4.2.2 Confidentiality

Confidentiality prevents sensitive information from reaching the wrong party, while making sure that the right party can in fact get it. So, while communicating the data in the network, no one can understood except intended recipient.

6.4.2.3 Availability

Availability requires that WSN assets, i.e., data, are available to authorized parties, i.e. CH and base station, at appropriate time and not prevented through this time. It is a requirement intended to assure that WSN work promptly and service is not denied to authorize parties when they request them. So, with availability services of a network should be available always even in presence of an internal or external attacks

6.4.2.4 Freshness

Freshness is a central goal which is violated by replay attacks in which the attacker retransmits an old data in order to waste the system resources or confusing the receiver, i.e., base station. Generally, it ensures that no old messages have been replayed.

6.4.2.5 Robustness

A secure clustering algorithm must be as robust as possible. The degree of robustness is measured by the count of attacks that the algorithm prevents. It also depends on the kind of attack, whether it is active or passive. The previous list of attacks is used here to evaluate the robustness of each of the secure clustering algorithms. In order to evaluate each of the existing clustering algorithms from the robustness point of view, we will use two notations: $P - R$ and $A - R$ to indicate its work against passive attack and active attack respectively.

6.4.2.6 Efficiency

Secure clustering algorithm must take into consideration the WSN resource limitations, i.e., sensor memory size, energy, and computation powers. That is refers to preventing the complex security procedures that may decrease the network lifetime. It must balance between the security issue and the network performance. This is refers to the efficiency of the secure clustering algorithm. We will evaluate efficiency the secure clustering algorithms using three criteria: required memory (M), energy consumption (E), and the processing time (P).

6.4.2.7 Dynamic Clustering

Dynamic clustering process aims to reforming the network structure after each round according to the updated status and characteristics of the sensor nodes, i.e., the remaining energy of each sensor. On the other side, the static clustering algorithm allows only the CH change after each round. It forms the network structure to a fixed set of clusters at the initial round and makes it unchangeable until the network become unavailable, i.e., all nodes consume their energy.

Therefore, we have to find simple solution that allow securing the dynamic cluster network while consuming as little energy as possible and is adapted to a low computing power. The remaining of this chapter discuss the existing schemes for secure clustering according to the previous criteria and proposes a complete security schema for routing data between sensors nodes, CHs, and the base station in cluster-based model for WSN.

Table 6.2 shows the notations for the previous criteria that we use to evaluate the secure clustering algorithms.

Table 6.2 The notations of the evaluation criteria for a secure routing protocol

Notation	Meaning
S-CH	Secure Cluster Head Selection
S-CF	Secure Cluster Formation
S-DA	Secure Data Aggregation
S-DR	Secure Data Routing
A_i	Attack identifier, i.e., A_1 means DOS attack
M	The required memory size
E	The energy consumption ratio
P	The required processing time/complexity of computations
D	Dynamic Clustering
S	Static Clustering

6.5 The Existing Schemes for Secure Clustering in WSN

In order to apply security for clustering model, many security procedures such as the data partitioning, using key management, intruder detection by location or trust management [56] have been proposed. Cryptographic techniques, such as encryption and hashing, are useful in addressing these concerns. However, the use of these schemes greatly increases the energy consumption of sensor nodes and thus shortens their lifetime [57] as they need Key management specially in case of using asymmetric key schema. In addition, most of the traditional key management schemes assume the relationship between nodes is fixed, while clusters as well as the relationship between nodes in hierarchical protocol are dynamic, so these schemes designed for flat networks need modifications to be applied for cluster-based WSNs [58]. Furthermore, in asymmetric key schema a larger sensor memory size is required for key storage.

On the other side, Key management scheme (specially symmetric key schema) has two main advantages: it is safer by realization of node-to-node authentication, and it saves energy which is a challenge for any secure protocol [59]. In order to make use of these advantages in clustering model, many dynamic key management techniques were proposed [60–64]. In these new schemes a Key is created for each cluster and it will be common among the cluster nodes to guarantee the confidential communication between them. After each round, the cluster key will changed with the changing of the CH. The main problem of these methods is its need for more computation and require more memory size to store the encryption Keys. These requirements affect directly the network lifetime. In the remaining of this section we discuss a list of the existing security solutions, they advantages and their limitations as the following.

6.5.1 Data Partitioning/Multi-path Routing

In this type of security schemes, the aim is to divide the information into several parts. If a sensor tries to send information, it cuts the data into several packets of fixed size. Each packet is sent on a different route. Packets pass in different nodes. When the packets are received by the sink, it brings them together to regenerate the original message. The main advantage of this method is that: the attacker has to catch all packets of a message if it wants to know the information. In order to do it, it has to be able to listen the entire network. It is more complicated for an attacker to have the information. On the other hand, this solution requires additional computations to collect the different packets to regenerate the message. In addition, it is not suitable for all cases of clustering model. It is also appropriate to the multi-hope clustering model in which a CH communicates with the base station through another CH. In most cases, data partitioning requires an additional security mechanism, i.e., cryptography, to protect the packets during transmission.

6.5.2 *Hashing*

Hash functions have a very simple purpose, they take a long message and generate a unique output value (called message digest) derived from the content of the message. Message digest can be generated by the sender and transmitted with the message to the receiver which uses the same hash function to recompute the digest. We can exploit the unique properties of hash function as: the input can be of any length, the output has a fixed length, the hash function is one-way, and the hash function is collision free to prevent the active attack that modifies and retransmits the message. In addition, most hash functions produce a 128-bit message digest which represents a solution of the memory size of the sensor nodes.

6.5.3 *Cryptography*

Due to the resource constraints of wireless sensors, public-key based cryptographic algorithms, i.e., RSA, are too complicated and energy-consuming for WSNs. However the symmetric cryptographic technique has its own qualities that always make it favorable as compared to public key cryptography for WSNs [65]. As a result, most of cryptography solutions in WSN use symmetric key for securing the network, which are more adapted, quicker to perform, and not consume more energy. Although the cryptography allows us to secure the confidentiality of data, its main problem is the key distribution, and we need to find an appropriate key management schema for the network.

According to [66], there are four types of key management techniques which can be used.

6.5.3.1 Global Key

In this method, one key is shared by the entire network. To send a message, information is encrypted with this key. Once the message is received, it can be decrypted with the same key. This solution is an energy-efficient solution of cryptography. The information is encrypted once by the sender and decrypted only once by the receiver. However, its the solution with a limited security. If an attacker could find the key, he is able to hear the entire network which communicates with this unique key. To know this key also allows the possibility to insert a malicious node in the network.

6.5.3.2 Pair Wise Key Node

Each node has a different key to communicate with a neighboring node which shares this key. So if one node has “n” neighbors, it has “n” key stored to communicate with its neighbors. In this solution, a node that sends a message has to encrypt the message with key neighbor who receives the information. The neighboring decrypts information to re-encrypt with the key corresponding to the following receiver. This solution increases considerably the security of the network, because if an attacker discovers a key, this key is just able to communicate with two nodes, and limits the power of this attack. The attacker has to find all pair wise key to listen the entire network. However, this technique is not energy-efficient especially in time of calculation, since each pair of nodes which transmits information has to encrypt and decrypt a message. The lifetime of the network and its rate is going to be reduced. So, we think it may be inefficient solution in case of clustering model because it will consume more energy from the CH in order to decrypt all messages from all sensors inside the cluster. Also, it requires additional memory size for the cluster head to store all keys of all nodes which will be impossible in case of dynamic model.

6.5.3.3 Pair Wise Key Group

Each group or cluster has a key to communicate between nodes in the cluster. This solution offers a compromise between security and energy efficiency. It may limit the number of encryption in communications. However it increases the work of clusters heads, which have to decrypt and encrypt the information. To be effective, we have to ensure that CHs change regularly in order not to consume all the energy of the CH. The main advantage of this method is that it can be applied to the dynamic clustering model.

6.5.3.4 Individual Key

In this solution, each node has its own key to encrypt data. This key is only known by the sink. As a consequence, a message sent by this node goes around hidden on the network until it reaches the sink. This solution is one of the better way to limit the consumption of the network. Nevertheless, this solution secures only communication between a node and the sink. In cluster model, this technique may consume the CH energy rapidly in case of many malicious node attached themselves to the cluster and sent unwanted messages to the CH. In such case, the CH will forward the data automatically to the base station without know its meaning. However, if we find a method to guarantee that the CH will know the source of the message, i.e. we can use the Node Coordinates as an identifier; this method can be used and represents a good solution.

6.5.4 Generation

Another key distribution solution is to use a key generation. Each round or generation, the sink sends a new key to the whole network. This key is used as a certificate for each node, to prove it belongs to the network. If an unidentified node tries to come into the wireless sensor network and if it does not have this key generation, the network will refuse its integration. Another benefit of this technique is that it limits substitution attacks of a sensor and the reprogramming of the sensor to be reused in the network. This technique is energy-efficient and easy to apply. However it directed only closed networks, which cannot accept new nodes. Moreover, there is the problem of a node, which cannot receive a key to progress time.

6.5.5 Localization

The work of this method is to use a technique for locating a node. For this solution, the wireless sensor network needs specific sensors called beacon node, which are sensors that knowing their geographical position. For example they can use a GPS equipment. The problem is that it cannot work on any other type of sensors.

6.5.6 Intrusion Detection System (IDS)

Intrusion is an unauthorized (unwanted) activity in a network that is either achieved passively (e.g., information gathering, eavesdropping) or actively (e.g., harmful packet forwarding, packet dropping, hole attacks). In a security system, if the first line of defense, Intrusion Prevention, does not prevent intrusions, then the second line of defense, Intrusion Detection, comes into play. It is the detection of any suspicious behavior in a network performed by the network members [51].

An IDS is also referred to as a second line of defense, which is used for intrusion detection only; that is, IDS can detect attacks but cannot prevent or respond. Once the attack is detected, the IDSs raise an alarm to inform the controller to take action [67].

6.6 Secure Clustering Algorithms

In all clustering methods, security and reliability aspects of clustering and cluster head election have gained modest attention so far. On the one hand, there are many papers that survey the security solutions applied in wireless sensor networks, e.g. [51, 53, 60, 66, 68–70]. These papers detail the common security issues in sensor networks, like authentication, intrusion detection, secure routing, secure data aggregation, etc. However, none of these papers address the issue of secure building and data transmission in particular.

On the other side, some papers, e.g. [71, 72], tackle the problem of secure clustering and secure CH election in sensor networks focusing on issues like dynamic key change, complexity, cluster head election criteria, and so on. Regrettably, the latter papers do not consider the security routing aspects of clustering [68]. In this section, we focus on the existing secure clustering algorithms for WSN as general to evaluate them according to the proposed criteria.

6.6.1 SLEACH

SLEACH protocol is the first attempt to build a secure version of the well known LEACH protocol. It is prevents sinkhole, selective forwarding and HELLO flooding attacks. SLEACH prevents an intruder node to send falsified data messages. But it doesn't guarantee confidentiality and availability. This algorithm works with homogeneous WSNs in which all nodes have the same characteristics, i.e., initial energy, and processing power. This algorithm make use of cryptography as the security mechanism by using symmetric-key methods. It can protect the network from outsider attack but it decreased the network efficiency and performance.

6.6.2 SS-LEACH

SS-LEACH [73] is another protocol that offers security while being energy efficient. For that, it works with multi-path CHs to communicate with the base station. To ensure security, it employs key pre-distribution and self-localization techniques. SS-LEACH is protected from selective forwarding, Hello flooding and Sybil attacks, but it controls neither data integrity nor freshness [60]. SSLEACH improves the network efficiency by improving the method of selecting CHs and forms dynamic multi-paths CHs chains to transfer data to the base station.

6.6.3 ESODR

In ESODR [74] method, each cluster is made up of a CH and multiple gateways (GWs) and other cluster members. ESODR combines hash function, symmetric key cryptographic algorithm, and public key cryptographic algorithm together. In ESODR, the computational complexity is low and has got good efficiency and scalability but it suffer from the dynamic clustering nature of the network. In addition, it requires more memory size to store both the encryption key and the hash digest.

6.6.4 SecLEACH

SecLEACH [61] is an improvement of SLEACH. It is a protocol for securing node-to-node communication in LEACH-based networks. It introduced symmetric key and one-way hash chain to provide different performance numbers on efficiency and security depending on its various parameter values. Although it provides authenticity, confidentiality, integrity and freshness for node-to-node communication, SecLEACH did not provide a solution for the compromised CH attack. This is because SecLEACH is vulnerable to key collision attacks and do not provide full connectivity.

6.6.5 RLEACH

RLEACH protocol attempts to apply random pairwise key (RPK) scheme onto LEACH. AS in LEACH, RLEACH operation is round based. It has three basic phases: shared-key discovery phase, cluster set-up phase and data transmission phase. RLEACH has the ability to resist to several attacks such as selective forwarding, sybil and hello flooding. Nevertheless, it is possible that an insider exercises sinkhole attack to be CH. Compromised node can also corrupt BS by the falsified data messages it sends [60].

6.6.6 ORLEACH

The same idea of RLEACH was applied by adding IDS mechanism as a new phase and produced a new method called ORLEACH [75]. ORLEACH operation is, therefore, divided into the following phases: Shared-key discovery phase, Cluster set-up phase, isolation of previously detected, attackers and MNs selection, Data transmission phase and Intrusion detection and alerting phase. Although this algorithm solved the problems of RLEACH specially whose are related to the active attacks, it is complexity increases the processing time and the consumed energy of the network which directly affect its efficiency.

6.6.7 NSKM

NSKM [76] is a secure clustering method that tries to solve the problems related to key management. It provides an efficient key distribution and establishment way by using three categories of keys; pre-deployed keys, network generated keys and the BS broadcasted keys. It works good against replay and node capture attacks. The selection of CH among nodes is based on its location and its distance to base station.

NSKM also ensures that the whole network is never compromised even if there has been an attack in the network by providing a secure data routing from CHs to the base station. Its main problem is it cannot work with dynamic clustering environment and suffers from active attacks, i.e., sinkhole and wormhole.

6.6.8 EECBKM

EECBKM [65] is a cluster based technique for key management which the clusters are formed in the network and the CHs are selected based on the energy cost, coverage and processing capacity. An EBS key set is assigned by the base station to every CH and cluster key to every cluster this proposed technique reduces node-capture attacks and efficiently increases packet delivery ratio with reduced energy consumption. But the problem of this protocol is that it works well in the environment with low density of sensors. In addition, it suffers many kinds of active attack. Another method is the SAC which is successful in preventing attacks caused by adversary like hello flooding and provides resilience to sensor nodes captured by adversary [58]. PIKE uses probabilistic techniques to establish pair wise keys between neighboring nodes in the network. However, in this approach, each node has to store a large number of keys.

6.6.9 SCMRP

Another secure clustering algorithm is SCMRP [77] which is based on multipath technique. SCMRP collects the benefits of both cluster based routing and multipath routing. It provides security against various attacks like altering the routing information, selective forwarding attack, sinkhole attack, wormhole attack, Sybil attack etc. In addition, it uses cryptography as a security mechanism to protect message after portioning it to packets. SCMRP consists of five phase; neighbor detection and topology construction, pairwise key distribution, cluster formation, data transmission, and re-clustering and rerouting. The Base station collects all the neighbor list from sensor node and apply an algorithm called DFS for finding multiple path. The BS generates the pairwise key and unicast to all nodes. The CH selection is based on the remaining energy of the node.

6.6.10 SHEER

SHEER [78] aims to create a secure clustering schema with energy-efficient and secure communication on the network layer. SHEER uses the cryptography as the security mechanism. It proposed a schema for key distribution based on the Hierarchical Key Establishment System (HKES). SHEER proposed also a probabilistic transmission mechanism to reduce energy consumption and extend the network lifetime. This method works effectively against HELLO flood attack, sybil attack and sinkhole attack. Its main drawback is that it is not able to protect the network from selective forwarding attacks.

6.6.11 AKM

AKM [79] is s cryptography-based method that provided security by using two kinds of keys: a pair-wise between the nodes inside the cluster, and a network key. This algorithm provides multiple level of encryption that works good with secure cluster formation and avoid node captures. AKM provides confidentiality, continuous authentication of nodes in the network by periodically changing the network key. However, if the compromised node attached with the network before refreshing the current network key, all the network operations of can be monitored.

6.6.12 SRPSN

SRPSN [80] is another cryptography-based in which a symmetric key is shared between all CHs and the base station to protect data. SRPSN dose not guarantee only the cluster building process, but also it is designed to protect the data packet transmission on the sensor networks under different types of attacks. Concerning to the key mechanism, this algorithm used the group key management scheme. However, one of its limitations is that there is no authentication in the mechanism. As a result, SRPSN fail to protect against many types of attacks specially spoofing, altering, replaying and sybil attack. Also, malicious node can also become a sinkhole.

6.6.13 SecRout

SecRout [81] aims to protect the network from compromised nodes attack. The main advantage of SecRout is its ability to detect the data modification if it occurs by malicious nodes during the transmission process. It uses efficient symmetric cryptography to secure data with two types of keys: the master shared key between the sink and CHs, and the cluster key among the clusters. Also, it guarantees freshness of data which enable it to catch any modified part. Another strength of SecRout is that it uses two-level architecture that reduces the communication overheads between nodes. Therefore, SecRout can greatly save the energy, and decrease the usage of memory and bandwidth.

6.6.14 IKDM

In IKDM [82] each node has a unique identifier (ID) in the network. It uses Pairwise key a mechanism for cryptography. The node ID is assigned at the initialization phase of the network by an offline Key Distribution Server (KDS). Then every nodes create a pair-wise key between them by exchanging their node IDs first. This method provides better network throughput and fixed key storage overhead and is suitable for large-scale WSNs. Therefore IKDM scheme is more energy-efficient due to the lower communication overhead for sensor nodes during the pair-wise key establishment process. Also, it can achieve better network resilience against node capture attack.

6.6.15 Genetic Algorithm-Based Techniques

In addition to the previous algorithms, a lot of security works based on intelligent techniques, i.e., GA, were proposed. For example, GBSWSHS was proposed at [63] to secure WSN which is used in health care applications [83–86]. In GBSWSHS method, the actual data is encrypted by using the key which is extracted from the receiver's fingerprint biometric. Second, to reduce a transmission-based attack, the fingerprint based cryptographic key is randomized by applying a genetic operator. However, the computation time, memory size, and the network lifetime are the main problems of this method. We will exclude GBSWSHS from our analysis because it was proposed as a general security method for WSN which was not created for the clustering model. Another GA based schema was proposed at [87]. This scheme is divided into three parts respectively for the base station, the CHs, and the sensor nodes. The base station first uses GAs to generate appropriate key-generating functions (KGFs) for re-keying on sensor nodes under energy consumption constraints. The functions are further divided into code slices which are then embedded into sensor nodes and headers before deployment. As sensor nodes are deployed, the CHs will randomly assemble the common slices and send the series to sensors for rebuilding the KGFs for re-keying. The re-keying functions are rebuilt in each predefined interval, such that it would be difficult for an attacker to crack the functions in time. But this method did not prevent the CH comprised attack is appropriate only for static clustering schema. The author of [62] proposed an IDS based on GA for detecting the misbehaviors based on node attributes. However, this algorithm applied to multilayer network such as multi-hope clustering model but the author did not provide additional information about the network structure building process. Finally, a novel artificial immune system based random keying technique for clustered sensor network was proposed on [59]. This algorithm works well with dynamic clustering environment. But according to [58], this scheme performs well against the outsider attack in comparison to the insider attack.

6.6.16 Additional Methods

In [88], a secure clustering method was proposed based on multi-path route discovering. This method was proposed to deal with the malicious behaviors of the data aggregation nodes and the malicious route behaviors of the nodes in WSN. In this method, the trusted value and residual energy for the nodes are used to choose the data aggregation nodes, a relatively reliable path is secretly selected to transfer the data aggregation results, and a secure clustering and reliable disjoint multi-path route discovery method is proposed by the functional-trust based secure data aggregation method. As general, this method represents an excellent solution for all types of passive attacks. It also provides a way to avoid the physical kinds of attacks like node destruction and node malfunction. On the other side, the efficiency is big challenge. A hybrid key management scheme for secure clustering in WSN was proposed at [89]. However, this method requires special characteristics for CH nodes. So, it works only with the heterogeneous clustering model in which nodes may differ in their features, i.e., processing power, memory size, initial energy, and transmission range.

6.7 Secure Clustering Algorithms Analysis and Evaluation

In this section, we provide a group evaluation of the discussed secure clustering protocols based on security goals, various routing attacks, performance, and cluster building metrics based on the proposed criteria. To clarify and summarize the advantages and the limitations of the above methods according to completeness, efficiency, and dynamic clustering criteria, Table 6.3 is constructed. Also, Table 6.4 provides the evaluation of these algorithms based on the robustness criteria and the security goals. Table 6.4 also provides a list of attacks that each protocol prevents.

Based on Tables 6.3 and 6.4, if we select the completeness as the only evaluation criteria, ORLEACH, SRPSN, SecRout, and the proposed algorithm in [88] are the most secure clustering algorithms. On the other side, the dynamic clustering criteria is applied by SLEACH, SecLEACH, RLEACH, ORLEACH, EECBKM, the proposed algorithm in [88], SHEER, AKM, SRPSN, SecRout, and IKDM algorithms. Where the efficiency criteria is applied by SecRout, AKM, and IKDM algorithms.

Related to the types of attacks, the work of most of the existing algorithms concentrated on preventing the CH attack in which an external malicious node tries to act as CH to collect data from the cluster members [69]. In addition, most of these procedures greatly decrease the network efficiency during the data aggregation process by using complicated cryptography schema. However, it seems that the algorithms that used cryptography and hashing together as the security mechanisms, i.e., ESODR and SecLEACH, are closer to the desired solution. Most of them provided a good solution for many kinds of both passive and active attack. We think that with searching about good solution for the key management problem, these algorithms

Table 6.3 Secure clustering protocols analysis according to completeness, efficiency, and dynamic clustering criteria

Algorithm	Mechanism	Completeness				Efficiency			Dynamic clustering	
		S-CH	S-CF	S-DA	S-DR	M	E	P	D	S
SLEACH	Cryptography	✓	✓	X	X	High	Low	Medium	✓	✓
ESODR	Cryptography + Hashing	✓	✓	X	✓	High	Low	Low	X	✓
SecLEACH	Cryptography + Hashing	✓	X	✓	✓	High	Low	Low	✓	✓
RLEACH	Cryptography	✓	✓	✓	X	Medium	Low	Medium	✓	✓
ORLEACH	IDS + Cryptography	✓	✓	✓	✓	High	High	Low	✓	✓
EECBKM	Cryptography	✓	✓	X	X	Low	Low	Low	✓	✓
Wang et al. [87]	Cryptography	X	✓	✓	✓	Medium	Medium	Low	X	✓
Radhika et al. [62]	IDS	–	–	–	–	Low	High	Medium	–	–
Sandeep et al. [59]	Cryptography	✓	✓	X	✓	Medium	Medium	Medium	✓	✓
NSKM	Cryptography	✓	✓	✓	X	Low	Medium	High	X	✓
SS-LEACH	Multi-path	✓	✓	X	X	High	Medium	High	X	✓
Zhong et al. [88]	Multi-path	✓	✓	✓	✓	Low	Medium	Medium	✓	✓
SCMRP	Multi-path	✓	✓	✓	✓	Medium	Medium	High	X	✓
Zhao et al. [89]	Cryptography	✓	✓	✓	X	High	High	Medium	✓	✓
SHEER	Cryptography	✓	✓	X	✓	Low	Low	Medium	✓	✓
AKM	Cryptography	✓	✓	✓	X	Low	Low	High	✓	✓
SRPSN	Cryptography	✓	✓	✓	✓	Medium	Medium	Medium	✓	✓
SecRout	Cryptography	✓	✓	✓	✓	Low	Low	High	✓	✓
IKDM	Cryptography	✓	X	✓	✓	Low	Low	High	✓	✓

Table 6.4 Secure clustering protocols with security goals and robustness criteria

Algorithm	Integrity	Confidentiality	Availability	Freshness	Prevents attacks
SLEACH	✓	X	✓	✓	$A_2, A_3, A_6, A_7, A_{11}, A_{14}$
ESODR	✓	✓	✓	X	$A_2, A_3, A_4, A_6, A_{10}, A_{11}, A_{14}$
SecLEACH	✓	✓	✓	✓	$A_2, A_4, A_6, A_7, A_{11}$
RLEACH	✓	X	X	✓	$A_2, A_3, A_4, A_6, A_{11}, A_{14}$
ORLEACH	✓	✓	✓	X	$A_1, A_2, A_3, A_4, A_6, A_9$
EECBKM	✓	✓	✓	✓	A_3, A_4, A_6, A_{12}
Wang et al. [87]	✓	X	✓	X	$A_2, A_3, A_4, A_5, A_6, A_{15}$
Radhika et al. [62]	✓	✓	X	✓	$A_1, A_2, A_6, A_7, A_8, A_{11}, A_{14}$
Sandeep et al. [59]	✓	X	X	✓	$A_2, A_3, A_4, A_6, A_7, A_{10}, A_{12}$
NSKM	✓	X	✓	✓	$A_2, A_3, A_4, A_5, A_6, A_{15}$
SS-LEACH	✓	X	✓	X	$A_2, A_4, A_6, A_7, A_9, A_{11}, A_{12}$
Zhong et al. [88]	✓	✓	✓	X	$A_1, A_2, A_9, A_{11}, A_{12}, A_{14}, A_{15}$
SCMRP	✓	✓	X	X	$A_2, A_3, A_5, A_6, A_7, A_{13}$
Zhao et al. [89]	✓	X	✓	X	$A_2, A_5, A_6, A_7, A_{11}, A_{13}, A_{14}$
SHEER	✓	✓	✓	✓	$A_3, A_4, A_5, A_6, A_9, A_{14}, A_{15}$
AKM	✓	✓	✓	✓	$A_2, A_4, A_5, A_6, A_7, A_{14}$
SRPSN	✓	X	✓	✓	$A_1, A_2, A_5, A_6, A_9, A_{11}, A_{13}$
SecRout	✓	X	✓	✓	$A_2, A_5, A_6, A_8, A_9, A_{10}, A_{13}, A_{15}$
IKDM	✓	✓	✓	✓	$A_2, A_3, A_4, A_6, A_9, A_{11}, A_{14}$

may achieve the required balancing between security requirements and the network performance.

Concerning to the security goals and robustness, we observe that AKM and IKDM secure clustering protocols maintain the most whereas SCMRP, RLEACH, and NSKM gain the least. We observe that SecLEACH, SHEER, EECBKM, AKM, and IKDM address all the security goals. Thus they are more secure than rest of the protocols if the security goals is taken as criteria. According to the security goals, all the listed secure clustering algorithms applied integrity.

Finally, we can say most of secure clustering protocols for WSNs use the symmetric key schemes due to their less computation time compared with the other schemes. Any secure clustering algorithm for WSN must guarantee not only the four phases of secure clustering, but also all other criteria which we used to evaluate the existing algorithms. For example, in ORLEACH algorithm, the four phases are applied but the algorithm still requires high memory storage for each sensor, consumes more energy through its need for additional processing and computation time. So, we cannot apply security and ignore the network performance which affects its lifetime.

6.8 Summary

Wireless Sensor Network (WSN) consists of set of sensor devices with limited resources and transpired in tangible insecure environments in order to collect data, which make security an essential challenge. For the sake of increasing the network lifetime and reducing the energy consumption, the cluster based model was proposed. In order to apply security for clustering model, many security procedures for the wireless sensor networks have been proposed. Most routing protocols are vulnerable to a number of security threats and are applied to the fixed clustering schema. Therefore, we review and critically discuss the most prominent secure clustering routing algorithms that have been developed for WSNs. Then, we explained the steps towards building a simple solution that allow securing the dynamic cluster network while consuming as little energy as possible and is adapted to a low computing power. Moreover, four phased towards building a secure clustering algorithm for WSNs are discussed. These phases are secure cluster head selection, secure cluster formation, secure data aggregation by the cluster head from its cluster nodes, and secure data routing to the base station. In order to build a secure clustering algorithm, this algorithm must guarantee not only the four phases of secure clustering, but also all our suggested criteria, i.e., efficiency, robustness, and dynamic clustering.

References

1. Yuan, X., Li, D., Mohapatra, D., & Elhoseny, M. (2017). Automatic removal of complex shadows from indoor videos using transfer learning and dynamic thresholding. *Computers and Electrical Engineering*. https://doi.org/10.1016/j.compeleceng.2017.12.026. (in Press).
2. Sajjad, M., Nasir, M., Muhammad, K., Khan, S., Jan, Z., Sangaiah, A. K., Elhoseny, M., & Baik, S. W. (2017). Raspberry Pi assisted face recognition framework for enhanced law-enforcement services in smart cities. *Future Generation Computer Systems*. Elsevier. https://doi.org/10.1016/j.future.2017.11.013.
3. Shehab, A., Elhoseny, M., El Aziz, M. A., & Hassanien, A. E. (2018). Efficient schemes for playout latency reduction in P2P-VoD systems. In A. Hassanien, & D. Oliva (Eds.), *Advances in soft computing and machine learning in image processing. Studies in computational intelligence*, Vol. 730. Cham: Springer. https://doi.org/10.1007/978-3-319-63754-9_22.
4. Elhoseny, M., Nabil, A., Hassanien, A. E., & Oliva, D. (2018). Hybrid rough neural network model for signature recognition. In A. Hassanien, & D. Oliva (Eds.), *Advances in soft computing and machine learning in image processing. Studies in computational intelligence*, Vol. 730. Cham: Springer. https://doi.org/10.1007/978-3-319-63754-9_14.
5. Abdeldaim, A. M., Sahlol, A. T., Elhoseny, M., & Hassanien, A. E. (2018). Computer-aided acute lymphoblastic Leukemia diagnosis system based on image analysis. In A. Hassanien, & D. Oliva (Eds.), *Advances in soft computing and machine learning in image processing. Studies in computational intelligence*, Vol. 730. Cham: Springer. https://doi.org/10.1007/978-3-319-63754-9.
6. Elhoseny, H., Elhoseny, M., Riad, A. M., & Hassanien, A. E. (2018). A framework for big data analysis in smart cities. In: A. Hassanien, M. Tolba, M. Elhoseny, & M. Mostafa (Eds.), *AMLTA 2018 the international conference on advanced machine learning technologies and applications (AMLTA2018)*, Advances in Intelligent Systems and Computing, Vol. 723. Cham: Springer. https://doi.org/10.1007/978-3-319-74690-6_40.

7. Elhoseny, M., Shehab, A., & Osman, L. (2018). An empirical analysis of user behavior for P2P IPTV workloads. In: A. Hassanien, M. Tolba, M. Elhoseny, & M. Mostafa (Eds.), *AMLTA 2018 the international conference on advanced machine learning technologies and applications (AMLTA2018)*, Advances in Intelligent Systems and Computing, Vol. 723. Cham: Springer. https://doi.org/10.1007/978-3-319-74690-6_25.
8. Wang, M. M., Qu, Z. G., & Elhoseny, M. (2017). Quantum secret sharing in noisy environment. In X. Sun, H. C. Chao, X. You, & E. Bertino (Eds.), *Cloud computing and security, ICCCS 2017*. Lecture Notes in Computer Science, Vol. 10603. Cham: Springer. https://doi.org/10.1007/978-3-319-68542-7_9.
9. Elsayed, W., Elhoseny, M., Riad, A. M., & Hassanien, A. E. (2018). Autonomic self-healing approach to eliminate hardware faults in wireless sensor networks. In A. Hassanien, K. Shaalan, T. Gaber, & M. Tolba (Eds.), *Proceedings of the international conference on advanced intelligent systems and informatics 2017, AISI 2017*, Advances in Intelligent Systems and Computing, Vol. 639. Cham: Springer. https://doi.org/10.1007/978-3-319-64861-3_14.
10. Shehab, A., Ismail, A., Osman, L., Elhoseny, M., & El-Henawy, I. M. (2017). Quantified self using IoT wearable devices. In A. Hassanien, K. Shaalan, T. Gaber, & M. Tolba (Eds.), *Proceedings of the international conference on advanced intelligent systems and informatics 2017, AISI 2017*, Advances in Intelligent Systems and Computing, Vol. 639. Cham: Springer. https://doi.org/10.1007/978-3-319-64861-3_77.
11. Elhoseny, M., Yuan, X., El-Minir, H. K., & Riad, A. M. (2016b). An energy efficient encryption method for secure dynamic WSN. *Security and Communication Networks*, *9*(13), 2024–2031.
12. Elhoseny, M., Elminir, H., Riad, A., & Yuan, X. (2016a). A secure data routing schema for WSN using elliptic curve cryptography and homomorphic encryption. *Journal of King Saud University-Computer and Information Sciences*, *28*(3), 262–275.
13. Elsayed, W., Elhoseny, M., Riad, A., & Hassanien, A. E. (2017). Autonomic self-healing approach to eliminate hardware faults in wireless sensor networks. In *International conference on advanced intelligent systems and informatics*, (pp. 151–160). Springer.
14. Elsayed, W., Elhoseny, M., Sabbeh, S., & Riad, A. (2017). Self-maintenance model for wireless sensor networks. *Computers and Electrical Engineering*. https://doi.org/10.1016/j.compeleceng.2017.12.022. (in Press).
15. Elhoseny, M., Yuan, X., El-Minir, H. K., & Riad, A. M. (2016). An energy efficient encryption method for secure dynamic WSN. *Security and Communication Networks*, *9*(13), 2024–2031. https://doi.org/10.1002/sec.1459.
16. Tharwat, A., Mahdi, H., Elhoseny, M., & Hassanien, A. E. (2018). Recognizing human activity in mobile crowdsensing environment using optimized k-NN algorithm. *Expert Systems With Applications*. https://doi.org/10.1016/j.eswa.2018.04.017. Accessed 12 April 2018.
17. Tharwat, A., Elhoseny, M., Hassanien, A. E., Gabel, T., & Kumar, A. (2018). Intelligent Bezir curve-based path planning model using chaotic particle swarm optimization algorithm. *Cluster Computing*, (pp. 1–22). Springer. https://doi.org/10.1007/s10586-018-2360-3.
18. Sarvaghad-Moghaddam, M., Orouji, A. A., Ramezani, Z., Elhoseny, M., & Farouk, A. (2018). Modelling the spice parameters of SOI MOSFET using a combinational algorithm. *Cluster Computing*. Springer. https://doi.org/10.1007/s10586-018-2289-6. (in Press).
19. Rizk-Allah, R. M., Hassanien, A. E., & Elhoseny, M. (2018). A multi-objective transportation model under neutrosophic environment. *Computers and Electrical Engineering*. Elsevier. https://doi.org/10.1016/j.compeleceng.2018.02.024.
20. Batle, J., Naseri, M., Ghoranneviss, M., Farouk, A., Alkhambashi, M., & Elhoseny, M. (2017). Shareability of correlations in multiqubit states: Optimization of nonlocal monogamy inequalities. *Physical Review A*, *95*(3), 032123. https://doi.org/10.1103/PhysRevA.95.032123.
21. Elhoseny, M., Yuan, X., El-Minir, H. K., & Riad, A. (2014). Extending self-organizing network availability using genetic algorithm. In *International conference on computing, communication and networking technologies (ICCCNT)*, (pp. 1–6). IEEE.
22. Elhoseny, M., Yuan, X., Yu, Z., Mao, C., El-Minir, H., & Riad, A. (2015). Balancing energy consumption in heterogeneous wireless sensor networks using genetic algorithm. *IEEE Communications Letters*, *19*(12), 2194–2197.

23. El Aziz, M. A., Hemdan, A. M., Ewees, A. A., Elhoseny, M., Shehab, A., Hassanien, A. E., & Xiong, S. (2017). Prediction of biochar yield using adaptive neuro-fuzzy inference system with particle swarm optimization. In *2017 IEEE PES PowerAfrica conference*, June 27–30, Accra-Ghana, pp. 115–120. IEEE. https://doi.org/10.1109/PowerAfrica.2017.7991209.
24. Ewees, A. A., El Aziz, M. A., & Elhoseny, M. (2017). Social-spider optimization algorithm for improving ANFIS to predict biochar yield. In *8th International conference on computing, communication and networking technologies (8ICCCNT)*, July 3–5. Delhi-India: IEEE.
25. Metawa, N., Elhoseny, M., Hassan, M. K., & Hassanien, A. E. (2016). Loan portfolio optimization using genetic algorithm: A case of credit constraints. In *Proceedings of 12th International Computer Engineering Conference (ICENCO)*, (pp. 59–64). IEEE. https://doi.org/10.1109/ICENCO.2016.7856446.
26. Elhoseny, M., Farouk, A., Zhou, N., Wang, M., Abdalla, S., & Batle, J. (2017a). Dynamic multi-hop clustering in a wireless sensor network: Performance improvement. *Wireless Personal Communications*, 1–21.
27. Elhoseny, M., Tharwat, A., Farouk, A., & Hassanien, A. E. (2017b). K-coverage model based on genetic algorithm to extend WSN lifetime. *IEEE Sensors Letters*, *1*(4), 1–4.
28. Elhoseny, M., Tharwat, A., Yuan, X., & Hassanien, A. E. (2018). Optimizing K-coverage of mobile WSNs. *Expert Systems with Applications*, *92*, 142–153. https://doi.org/10.1016/j.eswa.2017.09.008.
29. Elhoseny, M., Farouk, A., Batle, J., Shehab, A., & Hassanien, A. E. (2017). Secure image processing and transmission schema in cluster-based wireless sensor network. In *Handbook of research on machine learning innovations and trends*, Chapter 45, pp. 1022–1040, IGI Global. https://doi.org/10.4018/978-1-5225-2229-4.ch045.
30. Elhoseny, M., Elleithy, K., Elminir, H., Yuan, X., & Riad, A. (2015). Dynamic clustering of heterogeneous wireless sensor networks using a genetic algorithm towards balancing energy exhaustion. *International Journal of Scientific & Engineering Research*, *6*(8), 1243–1252.
31. Yuan, X., Elhoseny, M., El-Minir, H., & Riad, A. (2017). A genetic algorithm-based, dynamic clustering method towards improved WSN longevity. *Journal of Network and Systems Management*, *25*(1), 21–46.
32. Guo, W., & Zhang, W. (2014). A survey on intelligent routing protocols in wireless sensor networks. *Journal of Network and Computer Applications*, *38*, 185–201.
33. Ahmed, G., Khan, N. M., & Ramer, R. (2008). Cluster head selection using evolutionary computing in wireless sensor networks. In *Progress in electromagnetics research symposium*, (pp. 883–886).
34. Asim, M., & Mathur, V. (2013). Genetic algorithm based dynamic approach for routing protocols in mobile ad hoc networks. *Journal of Academia and Industrial Research*, *2*(7), 437–441.
35. Bhaskar, N., Subhabrata, B., & Soumen, P. (2010). Genetic algorithm based optimization of clustering in ad-hoc networks. *International Journal of Computer Science and Information Security*, *7*(1), 165–169.
36. Karimi, A., Abedini, S., Zarafshan, F., & Al-Haddad, S. (2013). Cluster head selection using fuzzy logic and chaotic based genetic algorithm in wireless sensor network. *Journal of Basic and Applied Scientific Research*, *3*(4), 694–703.
37. Hosseinabadi, A. A. R., Vahidi, J., Saemi, B., Sangaiah, A. K., & Elhoseny, M. (2018). Extended genetic algorithm for solving open-shop scheduling problem. *Soft Computing*. https://doi.org/10.1007/s00500-018-3177-y.
38. Rana, K., & Zaveri, M. (2013). Synthesized cluster head selection and routing for two tier wireless sensor network. *Journal of Computer Networks and Communications*, *13*(3).
39. Elhoseny, M., Ramírez-González, G., Abu-Elnasr, O. M., Shawkat, S. A., Arunkumar, N., & Farouk, A. (2018). Secure medical data transmission model for IoT-based healthcare systems. *IEEE AccessPP*(99). https://doi.org/10.1109/ACCESS.2018.2817615.
40. Shehab, A., Elhoseny, M., Muhammad, K., Sangaiah, A. K., Yang, P., Huang, H., & Hou, G. (2018). Secure and robust fragile watermarking scheme for medical images. *IEEE Access*, *6*(1), 10269–10278. https://doi.org/10.1109/ACCESS.2018.2799240.

41. Farouk, A., Batle, J., Elhoseny, M., Naseri, M., Lone, M., Fedorov, A., Alkhambashi, M., Ahmed, S. H., & Abdel-Aty, M. (2018). Robust general N user authentication scheme in a centralized quantum communication network via generalized GHZ states. *Frontiers of Physics, 13*, 130306. Springer. https://doi.org/10.1007/s11467-017-0717-3.
42. Elhoseny, M., Elkhateb, A., Sahlol, A., & Hassanien, A. E. (2018). Multimodal biometric personal identification and verification. In A. Hassanien, & D. Oliva (Eds.), *Advances in soft computing and machine learning in image processing. Studies in computational intelligence*, Vol. 730. Cham: Springer. https://doi.org/10.1007/978-3-319-63754-9_12.
43. Elhoseny, M., Essa, E., Elkhateb, A., Hassanien, A. E., & Hamad A. (2018). Cascade multimodal biometric system using fingerprint and Iris patterns. In A. Hassanien, K. Shaalan, T. Gaber, & M. Tolba (Eds.), *Proceedings of the international conference on advanced intelligent systems and informatics 2017, AISI 2017*, Advances in Intelligent Systems and Computing, Vol. 639. Cham: Springer. https://doi.org/10.1007/978-3-319-64861-3_55.
44. Elhoseny, M., Tharwat, A., & Hassanien, A. E. (2017c). Bezier curve based path planning in a dynamic field using modified genetic algorithm. *Journal of Computational Science*. https://doi.org/10.1016/j.jocs.2017.08.004.
45. Metawa, N., Hassan, M. K., & Elhoseny, M. (2017). Genetic algorithm based model for optimizing bank lending decisions. *Expert Systems with Applications, 80*, 75–82. https://doi.org/10.1016/j.eswa.2017.03.021.
46. Elhoseny, M., Shehab, A., & Yuan, X. (2017). Optimizing robot path in dynamic environments using genetic algorithm and Bezier curve. *Journal of Intelligent & Fuzzy Systems, 33*(4), 2305–2316. IOS-Press. https://doi.org/10.3233/JIFS-17348.
47. Attea, B. A., & Khalil, E. A. (2012). A new evolutionary based routing protocol for clustered heterogeneous wireless sensor networks. *Applied Soft Computing, 12*(7), 1950–1957.
48. Bayrakl, S., & Erdogan, S. (2012). Genetic algorithm based energy efficient clusters in wireless sensor networks. *Procedia Computer Science, 10*, 247–254.
49. Wu, Y., & Liu, W. (2013). Routing protocol based on genetic algorithm for energy harvesting-wireless sensor networks. *IET Wireless Sensor Systems, 3*(2), 112–118.
50. Hussain, S., Matin, A., & Islam, O. (2007). Genetic algorithm for energy efficient clusters in wireless sensor networks. In *International conference on information technology*.
51. Butun, I., Morgera, S., & Sankar, R. (2014). A survey of intrusion detection systems in wireless sensor networks. *IEEE Communications Surveys and Tutorials, 16*(1),
52. Padmavathi, G., & Shanmugapriya, D. (2009). A survey of attacks and security mechanisms and challenges in wireless sensor networks. *International Journal of Computer Science and Information Security, 4*(1).
53. Patel, M., & Aggarwal, A. (2013). Security attacks in wireless sensor networks: A survey. In *International conference on intelligent systems and signal processing*, (pp. 329–333).
54. Fuchsberger, A. (2005). Intrusion detection systems and intrusion prevention systems. *Elsevier Journal Information Security, 10*(3), 134–139.
55. Devi, C. D., & Santhi, B. (2013). Study on security protocols in wireless sensor networks. *International Journal of Engineering and Technology, 5*(5), 200–207.
56. Semary, A., & Abdel-Azim, M. (2013). New trends in secure routing protocols for wireless sensor networks. *International Journal of Distributed Sensor Networks*.
57. Zhang, M., Kermani, M., Raghunathan, A., & Jha, N. (2013). Energy-efficient and secure sensor data transmission using encompression. In *International conference on VLSI design and the 12th international conference on embedded systems*.
58. Singh, M., & Hussain, M. (2010). A top-down hierarchical multi-hop secure routing protocol for wireless sensor networks. *International Journal of Ad hoc and Sensor and Ubiquitous Computing, 1*(2).
59. Sandeep, E., Kusuma, S., & Kumar, B. (2014). A random key distribution based artificial immune system for security in wireless sensor networks. In *IEEE international students' conference on electronics, electrical and computer science*.
60. Grgic, K., Zagar, D., & Krizanovic, V. (2013). Security in IPv6-based wireless sensor network precision agriculture example. In *International conference on telecommunications*, (pp. 79–86).

61. Oliveira, L., Ferreira, A., Vilaca, M., Wong, H., Bern, M., Dahab, R., et al. (2007). Secleach-on the security of clustered sensor networks. *Signal Processing*, *87*(12), 2882–2895.
62. Radhika, B., Raja, P., Joseph, C., & Reji, M. (2013). Node attribute behavior based intrusion detection in sensor networks. *International Journal of Engineering and Technology*, *5*(5), 3692–3698.
63. Shanthini, B., & Swamynathan, S. (2012). Genetic- based biometric security system for wireless-sensor-based healthcare systems. In *International conference on recent advances in computing and software systems*, (pp. 180–184).
64. Sharma, S., & Jena, S. (2011). A survey on secure hierarchical routing protocols in wireless sensor networks. In *International conference on communication, computing and security*.
65. Lalitha, T., & Umarani, R. (2012). Energy efficient cluster based key management technique for wireless sensor network. *International Journal of Advances in Engineering and Technology*, *3*(2), 186–190.
66. Schaffer, P., Farkas, K., HorvTh, A., Holczer, T., & ButtyN, L. (2012). Secure and reliable clustering in wireless sensor networks: A critical survey. *The International Journal of Computer and Telecommunications Networking*, *56*(11), 2726–2741.
67. Alrajeh, N., Khan, S., & Shams, B. (2013b). Intrusion detection systems in wireless sensor networks: A review. *International Journal of Distributed Sensor Networks*.
68. Alrajeh, N., Khan, S., Lloret, J., & Loo, J. (2013a). Secure routing protocol using cross layer design and energy harvesting in wireless sensor networks. *International Journal of Distributed Sensor Networks*.
69. Diop, A., Qi, Y., Wang, Q., & Hussain, S. (2013). An advanced survey on secure energy efficient hierarchical routing protocols in wireless sensor networks. *International Journal of Computer Science Issues*, *10*(1).
70. Salehi, A., Razzaque, M., Naraei, P., & Farrokhtala, A. (2013). Security in wireless sensor networks: Issues and challanges. In *IEEE international conference on space science and communication*, (pp. 356–360).
71. Rifa-Pous, H., & Herrera-Joancomart, J. (2011). A fair and secure cluster formation process for ad hoc networks. *Wireless Personal Communications*, *56*(3), 625–636.
72. Wang, G., Kim, D., & Cho, G. (2012). A secure cluster formation scheme in wireless sensor networks. *International Journal of Distributed Sensor Networks*.
73. Wu, D., Hu, G., & Ni, G. (2008). Research and improve on secure routing protocols in wireless sensor networks. In *Fourth IEEE international conference on circuits and systems for communications*, (pp. 853–856). IEEE.
74. Zhang, Y., & Xu, L. (2008). An efficient secure on-demand routing in clustered wireless ad hoc networks. In *International conference on wireless communications, networking and mobile computing*. IEEE.
75. Sahraoui, S., & Bouam, S. (2013). Secure routing optimization in hierarchical cluster-based wireless sensor networks. *International Journal of Communication Networks and Information Security*, *5*(3).
76. Gawdan, I., Chow, C., Zia, T., & Sarhan, Q. (2011). A novel secure key management for hierarchical wireless sensor network. In *Third conference on computational intelligence, modeling and simulation*, (pp. 312–316). IEEE.
77. Kumar, S., & Jena, S. (2010). SCMRP: secure cluster based multipath routing protocol for wireless sensor networks. In *Sixth international conference on wireless communication and sensor networks*, (pp. 1–6). IEEE.
78. Ibriq, J., & Mahgoub, I. (2006). A secure hierarchical routing protocol for wireless sensor networks. In *IEEE International conference on communication systems*, (pp. 1–6). IEEE.
79. Kausar, F., Masood, A., & Hussain, S. (2008). An authenticated key management scheme for hierarchical wireless sensor work. *Advances in Communication Systems and Electrical Engineering*, *4*, 85–98.
80. Tubaishat, M., Yin, J., Panja, B., & Madria, S. (2004). A secure hierarchical model for sensor network. *ACM Sigmod Record*, *33*(1), 7–13.

81. Yin, J., & Madria, S. (2006). Secrout a secure routing protocol for sensor network. In *IEEE International conference on advanced information networking and applications*, Vol. 1.
82. Cheng, Y., & Agrawal, D. (2007). An improved key distribution mechanism for large scale hierarchical wireless sensor networks. *Ad Hoc Networks*, *5*(1), 35–48.
83. Elhoseny, M., Abdelaziz, A., Salama, A. S., Riad, A. M., Muhammad, K., & Sangaiah, A. K. (2018). A hybrid model of internet of things and cloud computing to manage big data in health services applications. *Future Generation Computer Systems*. Elsevier. (in Press).
84. Abdelaziz, A., Elhoseny, M., Salama, A. S., & Riad, A. M. (2018). A machine learning model for improving healthcare services on cloud computing environment. *Measurement*, *119*, 117–128. https://doi.org/10.1016/j.measurement.2018.01.022.
85. Darwish, A., Hassanien, A. E., Elhoseny, M., Sangaiah, A. K., & Muhammad, K. (2017). The impact of the hybrid platform of internet of things and cloud computing on healthcare systems: Opportunities, challenges, and open problems. *Journal of Ambient Intelligence and Humanized Computing*. Springer. https://doi.org/10.1007/s12652-017-0659-1.
86. Abdelaziz, A., Elhoseny, M., Salama, A. S., Riad, A. M., & Hassanien, A. E. (2018). Intelligent algorithms for optimal selection of virtual machine in cloud environment, towards enhance healthcare services. In A. Hassanien, K. Shaalan, T. Gaber, & M. Tolba (Eds.), *Proceedings of the international conference on advanced intelligent systems and informatics 2017, AISI 2017*, Advances in Intelligent Systems and Computing, Vol. 639. Cham: Springer. https://doi.org/10.1007/978-3-319-64861-3_27.
87. Wang, L., Wang, C., & Liu, C. (2009). Optimal number of clusters in dense wireless sensor networks: A cross-layer approach. *IEEE Transactions on Vehicular Technology*, *58*(2), 966–976.
88. Zhong, C., Yinghong, M., Zhao, J., Lin, C., & Lu, X. (2014). Secure clustering and reliable multipath route discovering in wireless sensor networks. In *Sixth international symposium on parallel architectures and algorithms and programming*, pp. 130–134. IEEE.
89. Zhao, P., Xu, Y., & Nan, M. (2012). A hybrid key management scheme based on clustered wireless sensor networks. *Wireless Sensor Network*, *4*, 197–201.

Chapter 7
An Encryption Model for Data Processing in WSN

Abstract Building a secure routing protocol in WSN is not trivial process. It looks like an optimization process through which we try to find the optimum solution that maximize the network performance in an environment with a set of complicated constraints. The main purpose is not only to design new routing protocol that guarantee the network efficiency, but also balancing between this efficiency and the security requirements. For that purpose, we designed and followed a general framework to simplify the process of building such that protocol. In this chapter, an overview of the working steps towards building the proposed protocol is described. Then the protocol objectives and methodology are discussed.

7.1 Introduction

This chapter proposes a novel encryption schema based on genetic operations, i.e., crossover, and mutation [4]; to secure data transmission in WSN with dynamic clustering nature. There are two main purposed of our proposed schema. First, it aims to protect the network from many types of attack [5–9] such as passive attack, node computerized attack, CH compromised attack, and brute force attack [10–14]. The second purpose is to maximize the network performance by extending the network lifetime in both static and dynamic cluster-based WSN [15, 15–18], reducing the consumed energy specially for cluster head nodes, and shrinking the required memory size that is required for encryption key management. The proposed size of the encryption key at each sensor node is 176-bits which is composed of a combination of static and dynamic parameters, i.e., node ID, distance between node and its CH, and round index, to avoid active attack. To avoid the network communication overload, the authentication process between any two parities is built upon the first part of the key which is produced by Elliptic Curve Cryptography. To prevent its energy consumption, CH aggregates the encrypted data of its cluster members without having to decrypt them using the addition operation of Homomorphic Encryption.

Figure 7.1 shows the phases of building a secure routing protocol in details. It contains the main phases that was followed to create our secure routing protocol for WSN. This figure organizes the main steps and their relationship. There are three main

M. Elhoseny and A. E. Hassanien, *Dynamic Wireless Sensor Networks*, Studies in Systems, Decision and Control 165, https://doi.org/10.1007/978-3-319-92807-4_7

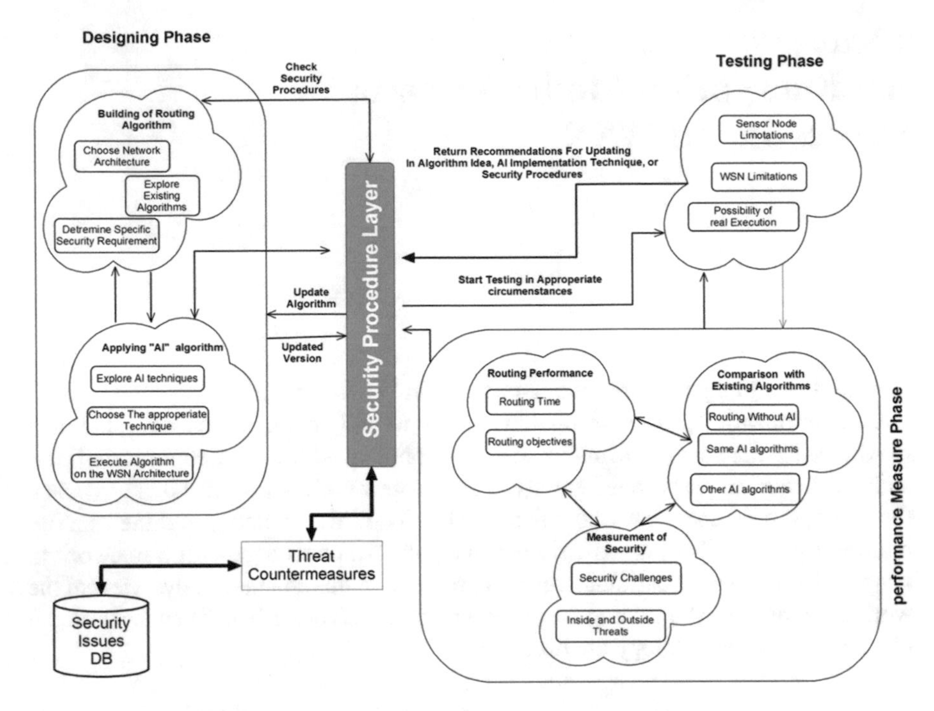

Fig. 7.1 Phases of building a secure routing protocol

phases: Design phase, testing phase, and performance measure phase. In addition, and intermediate layer called ‘Security Procedure Layer (SPL)’ is proposed to work as a reference for all other phases as described later. Each of these steps contains a set of tasks and is directly related to the remaining phases. Each phase is explained in details as the following.

7.1.1 Designing Phase

The design phase aims to set the main guidelines towards the actual implementation of GASONeC. These guidelines can be determined by following two main tasks. First, determine the main requirements towards building a secure routing algorithm. In order to do that, this task contains three sub-tasks:

- **Choose Network Architecture**: During this task, we explored the existing network structures, such as flat-routing and cluster-routing, and determined the strengthens and limitations of each. A comparison was conducted between all of them and based on it we decided to work with clustering model [19, 20].

- **Explore Existing Routing Algorithms**: The goal of this subtask is to literature the existing algorithms to define their limitations. In addition, a proposed solution for each limitation was provided.

- **Determine Specific Security Requirement**: The security requirements are determined based on the traditional and advanced limitations of WSN. A set of attacks and threats were defined in this subtask. The aim of this subtask is to clarify the threats and limitations towards building a secure routing protocol in WSN.

The second task in the design phase is responsible of choosing the appropriate intelligent technique of building the algorithm. This tasks includes three main subtasks as the following:

- **Explore Intelligent Techniques**: There are a lot of intelligent techniques that are proposed to constructing and organizing WSN's operations [21]. Various researches [21–25] discussed the routing protocols in Cluster-base WSN based on intelligent algorithms as reinforcement learning (RL), ant colony optimization (ACO), fuzzy logic (FL), genetic algorithm (GA) [26–28], and neural networks (NNs). Moreover, Many clustering methods have been proposed. Local Negotiated Clustering Algorithm (LNCA) employs the similarity of data acquired by nodes as a clustering criterion. ACE uses node degree as a feature to group sensor nodes into clusters. GA-WCA uses a load balanced factor and total distance from the neighboring nodes. LA2D-GA takes distance as the only parameter to calculate the fitness; however, a chromosome is represented as a two-dimensional grid [29]. Based on that, a comparison was conducted between the existing methods to evaluate the network performance in terms of security and network lifetime.

- **Choose the Appropriate Technique**: This sub-task aims to select the most appropriate technique that matches the requirements and limitations of WSN based on the conclusion of the previous sub-task. As mentioned on literature review at Chap. 3, we decided to use Genetic Algorithm (GA). GA has been applied in the routing protocol of WSN [23, 30–32]. A key objective is to define an appropriate fitness function that encodes the network structure and its goodness. In most of researches that uses GA to manage WSN, the main objective is to select the CHs with assumption that the number of clusters is determined in advance. in addition to that, the process of CH selection depends on the following factors: [22, 23, 25, 33] distance of a node from the cluster centroid, the remaining battery power, the degree of mobility, and the vulnerability index. While in other recent researches such as [24] less criteria are considered for cluster head selection.

- **Implement the Proposed Method Using the Selected Technique**: At this point, the problem is defined and the implementation technique is clarified. Based on that, the proposed protocol is developed and evaluated using the SPL. First, the dynamic clustering parameters are proposed. Then, GA is used to construct the network as will be discussed later. After that, the cryptography schema is applied to protect the data routing process. The SPL is used to measure if the implementation process matches the security goals.

7.1.2 Testing Phase

After finishing the implementation of the proposed protocol, the testing phase starts. It aims to evaluate the implementation details of the proposed algorithm to explore the implementation errors. Based on the individual testing of each task at the design phase, this phase executes complete test to evaluate the algorithm work using the WSN limitations. In addition, it measures to what extend the proposed algorithm can be applied in the real world. Similar to the individual test of each sub-task in the design phase, the testing process consults the SPL to make sure that the algorithm provide the acceptable limit of protection against the pre-defined attacks.

7.1.3 Performance Measure Phase

Secure data transmission and extending WSN longevity are our two main goals. In order to build a secure routing protocol that extend the network lifetime, a set of criteria and conditions have to be achieved as described at Chaps. 1 and 2. The goal of this phase is to evaluate and compare the performance of the proposed protocol with the existing state-of-the art methods using the same parameters and experimental environment. In addition, the protocol efficiency is measured. This phase contains three main tasks:

- **Comparison with Existing Protocols**: This task aims to conduct a comparison with the existing algorithms that were discussed as Chap. 2. The comparison includes both traditional and intelligent algorithms. The evaluation process include a set of points such as the energy consumption, memory requirement, and network lifetime.

- **Measurement of Security Goals**: The goal of this subtask is to test the performance by simulating the work of each attack. This sub-task aims to find the weakness of our protocol to be taken into consideration. Moreover, it tries to evaluate the performance of the network in different environments using different types of threats.

- **Routing Performance**: As we mentioned before, the efficiency is an important evaluation criteria for any protocol. This sub-task evaluates the efficiency of our protocol using the round time. In addition, it tries to find the change of the efficiency in different environments using different data, such as text and images.

7.1.4 *Security Procedure Layer*

During each task, SPL represents the guide to determine whether the proposed idea follows the security requirements or not. This process is done using a set of predefined attacks and their countermeasures stored at the security database. During the design phase, SPL is used to check and update the work while in testing and performance measure phases it aims to evaluate the performance of the protocol against a set of attacks.

7.2 Objectives and Contributions

Despite the great efforts in secure clustering of WSN, the dynamic nature and limited resources of SNs make searching for a secure and optimal network structure an open challenge. That is due to the characteristics of WSN, such as open communication medium, limited computational capabilities of nodes, and the disadvantages of bandwidth constraint [34], which make them more susceptible to malicious attacks than other networks. As a result, public-key based cryptographic algorithms, i.e., RSA, are too complicated and energy-consuming for WSNs. However the symmetric cryptographic technique has its own qualities that always make it favorable as compared to public key cryptography for WSNs [35]. based on that, most of cryptography solutions in WSN use symmetric key for securing the network, which are more adapted, quicker to perform, and not consume more energy. Although the cryptography allows us to secure the confidentiality of data, its main problem is the key distribution, and we need to find an appropriate key management schema for the network. In addition, CH in the network that responsible of the data aggregation from a set of SNs consumes more energy than common nodes and will quit the mission in advance due to energy exhausting because it must decrypt the data first and re-encrypt them after aggregating the data, moreover, it will bring complex key management to ensure the security of corresponding keys. WSNs thus require efficient encryption schemes in terms of memory size, energy consumption, and operating speed.

To address these challenges, a lot of work e.g. [36, 37] tackle the problem of secure clustering in sensor networks focusing on issues like dynamic key change, complexity, CH election criteria, and so on. For example, SLEACH [38] is the first attempt to build a secure clustering model. It is prevents sinkhole, selective forwarding and HELLO flooding attacks. However, traditional encryption-based methods in general and SLEACH in particular are suffer from the required memory size as well as a problems related to network performance and lifetime. To solve these problems, Elliptic Curve Cryptography (ECC) is used due to its ability to provide high security with short key size. In addition, the homomorphic encryption schema [39] is used to allow CHs to aggregate encrypted messages directly from sensors without decrypting so that it has a short aggregation delay. Moreover, it is prevent the attacker to know anything even if the CH is compromised, because CH is not responsible to encrypt messages. For example, TinyPEDS [40] was proposed as novel scheme for secure data aggregation at the CH based on privacy homomorphic encryption. However, this schema cannot resist node compromise attacks [39]. Similar to TinyPEDS, SEDA-ECC [39] was proposed based on the principles of privacy homomorphic encryption. The security results of this method is better than TinyPEDS specially in node compromise attack. But, the required memory size and the energy consumption represent the main challenges.

The contribution of this work is two-fold:

First, preventing most of cluster-based attacks, i.e. CH compromised, and node compromised attack; by diversification the encryption parameters every round. This is done by combining static parameters stored in the internal memory of node, i.e. ECC parameters; with dynamic parameters that can be calculated during the run time of the network operations, i.e. round index, and distance between the node and its cluster head. In addition, the proposed schema avoids the active attacks that aim to affect the data integrity during the transmission phase. This is because the ciphertext is based on the current state of the whole network instead of a set of static parameters.

Second, increasing the network performance by avoiding the high energy consumption of the CH through data aggregation process. The main advantage of the proposed scheme is its ability to deal with low sensors resources, i.e., memory size and processing power.

The rest of this article is organized as follows: Section 7.3 presents the related work of constructing secure clusters to extend network life and avoid a specific kinds of attacks on WSN. Section 7.4 describes our proposed secure dynamic clustering scheme. Section 7.5 discusses our experimental results in comparison to the state-of-the-art methods. Section 7.6 concludes this article.

7.3 Related Work

SLEACH [38] protocol is the first attempt to build a secure version of the well-known LEACH protocol. It is prevents sinkhole, selective forwarding and HELLO flooding attacks. SLEACH prevents an intruder node to send falsified data messages. But it doesn't guarantee confidentiality and availability. Based on SLEACH, SecLEACH [41] was proposed as an improvement to introduce symmetric key and one-way hash chain. The main aim of that is to provide different performance numbers on efficiency and security depending on its various parameter values. Although it provides authenticity, confidentiality, integrity and freshness for node-to-node communication, SecLEACH did not provide a solution for the compromised CH attack. This is because SecLEACH is vulnerable to key collision attacks and do not provide full connectivity. Furthermore, TinyPEDS [40] was proposed as novel scheme for secure data aggregation at the CH based on privacy homomorphic encryption. However, this schema cannot resist node compromise attacks [39].

SEDA-ECC [39] was proposed as a Secure-Enhanced Data Aggregation based on Elliptic Curve Cryptography. Similar to TinyPEDS, The design of SEDA-ECC is based on the principles of privacy homomorphic encryption. The security results of this method is better than TinyPEDS specially in node compromise attack. But, the required memory size and the energy consumption represent the main challenges. Another method based on the identity digital signature (IBS) scheme called SET-IBS [37] was proposed for secure and efficient data transmission for cluster based WSN. In SET-IBS, security relies on the hardness of the Diffie-Hellman problem in the pairing domain. It provides a good solution for both active and passive attacks but the network lifetime still represent the main problem. Due to large computational overhead of the asymmetric cryptography, BGN [42] was constructed on a cyclic group of elliptic curve point. In [43], several public-key-based encryptions schemes, i.e., EC-OU; were proposed to achieve data concealment in WSNs. A Well-known encryption schema is the Advanced Encryption System (AES). This algorithm is based on block cipher encryption. It has a fixed block size of 128 bits and has a key size of 128, 192, or 256 bits. However, AES running on 10, 12, and 14 rounds for 128, 192, and 256-bits key respectively is still found vulnerable by the researchers [44]. In addition, AES is not suitable for WSNs as it needs more hardware resources [45].

Another widely used block cipher schema in WSN is Skipjack [46]. It uses an 80-bits key to encrypt or decrypt 64-bit data blocks. This makes it vulnerable to the key search attack due to the short key length [47]. However, the strategy is the main challenge of SkipJack in WSNs. To avoid the short key problem of Skipjack, TWINE [48] uses an 80 bits or 128 bits key. The best known attacks against TWINE are two biclique attacks with a data requirement for the two attacks equal to 260 [49]. Based on lightweight block cipher schema, LED [50] was proposed to encrypts 64 bits blocks using either 64 bits or 128 bits key. To generate the new key in LED, the key is XORed at every four rounds. LED has two main limitations, it cannot prevent man-in-the middle attack and consumes more CPU cycles compared to conventional cryptographic schemes [51].

BCC [52] is a block cipher that avoids floating-point operations and multiplications in order to minimize the energy consumption. It is a flexible encryption method that supports different number of rounds and length of the encryption key. But, the authors failed to present any security and performance analysis for it. In addition, an encryption schema based on EEC and Chaotic Map was proposed in [51]. The proposed cryptographic scheme employs elliptic curve points to verify the communicating nodes and as one of the chaotic map parameters to generate the pseudorandom bit sequence. This sequence is used in XOR, mutation and crossover operations in order to encrypt the data blocks. However, the memory size is the big challenge of this schema. SecDAO-LEACH [53] was proposed to enhance the WSN performance in terms of security, reliability and fault-tolerance. This protocol is resistant to security attacks such as, replay attacks, node compromising attacks and impersonation attacks. In addition, it performs better in terms of energy consumption. However, it is suffer from the limited resources of sensors specially the required memory size.

EECBKM [35] is a cluster based technique for key management which the clusters are formed in the network and the CHs are selected based on the energy cost, coverage and processing capacity. An EBS key set is assigned by the base station to every CH and cluster key to every cluster this proposed technique reduces node-capture attacks and efficiently increases packet delivery ratio with reduced energy consumption. But the problem of this protocol is that it works well in the environment with low density of sensors. In addition, it suffers many kinds of active attack. Another method is the SAC which is successful in preventing attacks caused by adversary like hello flooding and provides resilience to sensor nodes captured by adversary [54]. PIKE uses probabilistic techniques to establish pair wise keys between neighboring nodes in the network. However, in this approach, each node has to store a large number of keys.

7.4 The Proposed Secure Clustering Schema

Our method is built upon GASONeC [55], which employs Genetic Algorithm (GA) to optimize sensor clusters for balancing energy consumption among sensors and to extend network life. In GASONeC, factors such as the remaining energy, the expected energy expenditure, the distance to the base station, and the number of nodes in the vicinity are used to search for an optimal [56–62], dynamic network structure, which are encoded into the fitness function of GA. In the optimization process, each GA chromosome represents a designation map of cluster heads. A gene in a chromosome specifies if the corresponding node serves as a cluster head. Given a cluster head, clusters of nodes are formed following the nearest neighbor rule, and the fitness of a WSN structure encoded in the chromosome is hence determined by the evaluation of all sensors alive. In each transmission round, network structure is updated according to the optimal balancing of the above factors to extend the network longevity. To ensure security in such a dynamically clustered sensor network, the security protocol must also take energy consumption into consideration. This secure schema enables a WSN to work with varity of recent applications that need high security requirements [63–71].

7.4.1 Key Establishment and Plaintext Encryption

A new encryption key is created in each transmission round for each sensor node. To overcome memory constraint, we design the encryption key of 176 bit that include the following components. The first component consists of 128 bits produced by the ECC and the ECC parameters are embedded in the sensors. The second component of the key represent the node ID in 13 bits. Although a typical size of a WSN is in the range of 200 nodes in most practical applications [37, 72–77], a large scale WSN could consist of thousands of nodes. Hence, we allocate 13 bits for the node ID that is capable to represent more than eight thousand sensors. The third component consists of 15 bits to encode the distance between the sensor and its CH $D_{i,CH}$, which is

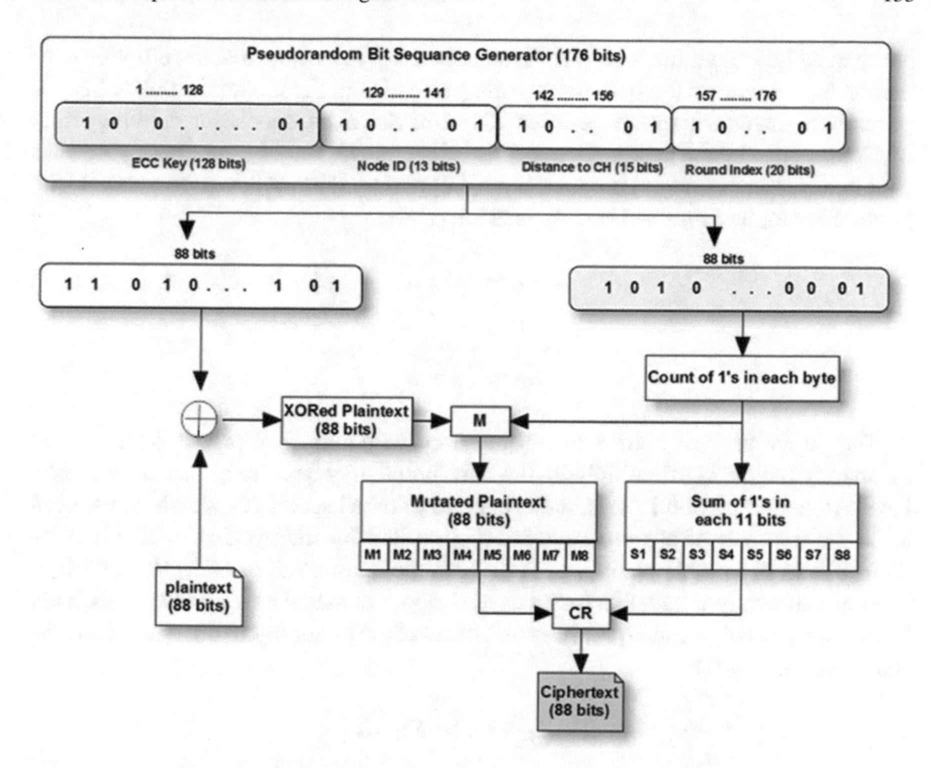

Fig. 7.2 The proposed encryption scheme

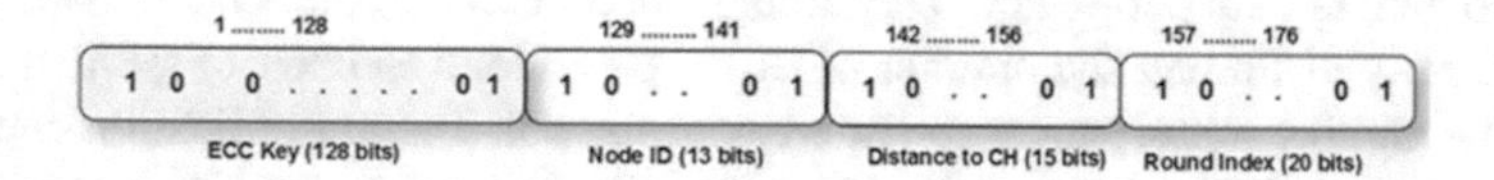

Fig. 7.3 The encryption key

followed by 20 bits to represent the index of transmission rounds R_t. An illustration of a key is shown in Fig. 7.3. The key establishment between a SN and the BS is a two-party authentication process. The cluster heads (CH) send the hashed ECC key from the member nodes to the BS to initialize the transmission sessions. We assume that every node knows its ID. The hashed code of the ECC key is produced using SHA-2 hashing function [39].

To be self-content, we briefly review ECC encryption. Generally, an elliptic curve is a cubic equation with the following form:

$$y^2 = x^3 + ax + b, \tag{7.1}$$

where a and b are integers that satisfy $4a^3 + 27b^2 \neq 0$. The first step before data transmission between SNs, an elliptic curve and a base point p that lies on the curve must be known for every SN in the network. We assume that the elliptic curve parameters as well as the base point p are preloaded in the memory of each SN. To generate a shared secret key between SN A to SN B, A chooses a random prime integer k_A and B chooses a random prime integer k_B. k_A and k_B are the private keys of A and B, respectively. After that, Eqs. (7.2) and (7.3) are followed to generate the public keys $\bar{k_A}$ and $\bar{k_B}$ for A and B, respectively.

$$\bar{k_A} = k_A * p \tag{7.2}$$

$$\bar{k_B} = k_B * p \tag{7.3}$$

The public keys of both A and B are a curve points. The private keys k_A and k_B specify the times of multiplying the base point p by itself to generate the public keys. After sharing public keys, they generate a shared secret key R, which we need to be the first part of our proposed encryption key, by multiplying public keys by their private keys as shown in Eq. (7.4). With the known values of $\bar{k_B}$, $\bar{k_A}$, and p, it is computationally intractable for an eavesdropper to calculate k_A and k_B which are the private keys of A and B. As a result, adversaries cannot figure out R which is the shared secret key [78].

$$R = k_A * \bar{k_B} = K_B * \bar{k_A} \tag{7.4}$$

The SN uses the main operations of point addition and point doubling of the elliptic curve to generate its public key. The strength of an ECC crypto-system depends on the difficulty of finding the number of times p is added to itself to get this public key. This number represents the private key of the SN. The multiplication operation can be executed by adding a point along an elliptic curve to itself repeatedly. The addition operation for any two points $\omega(x_1, y_1)$ and $\vartheta(x_2, y_2)$ over an elliptic curve is given by Eqs. (7.5) and (7.6) with the assumption that $\omega + \vartheta$ is equal to a new point $\theta(x_3, y_3)$. Figure 7.4 is an example of adding two points p_1 and p_2 on the curve.

$$x_3 = \lambda^2 - x_1 - x_2 \tag{7.5}$$

$$y_3 = \lambda * (x_1 - x_3) - y_1 \tag{7.6}$$

where λ is calculated as the following:

$$\lambda = \begin{cases} \frac{y_2 - y_1}{x_2 - x_1} & \text{if } \omega \neq \vartheta \\ \frac{3x_1^2 + a}{2y_1} & \text{if } \omega = \vartheta \end{cases}$$

Figure 7.2 shows the main steps of the encryption process. The processes M and CR denote permutation and concatenation operations respectively. The proposed cryptographic scheme implements three different operations: XOR, permutation,

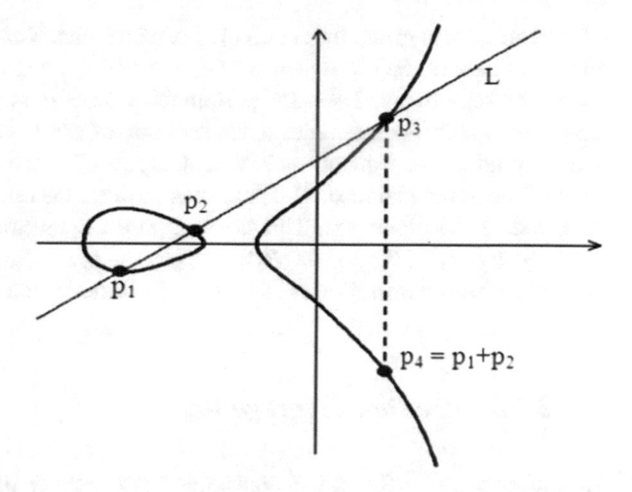

Fig. 7.4 Point addition in ECC

and concatenation. The additive cipher XOR is secure when the keystream is as long as the plaintext [51]. Permutation is a process of changing one or multiple bits value in a given bit string. Concatenation is a process of taking two parent bit strings and producing corresponding child bit strings by interchanging selected parts of bit strings between the parents. The permutation process is applied to create random diversity in the ciphertext, whereas the concatenation operation is used to change the order of the mutated text or image data. The main benefit of using these two operations is that they introduce relatively fair diversity in the ciphertext.

Algorithm 5: Plaintext Encryption Process

1: Partition the plaintext into 88-bits N blocks $S = \{b_1, b_2, \ldots, b_N\}$.
2: Generate the 176-bits bit sequence β
3: Divide β to two equal parts p_1 and p_2, 88-bits each.
4: Calculate the count of 1's for each byte $X[x_1, x_2, \ldots, x_8]$ as well as count of 1's for each 11 bits $Y[y_1, y_2, \ldots, y_8]$ in p_2
5: **for** $i = 1, 2, \ldots, N$ **do**
6: $\delta = p_1 \oplus b_i$
7: $\bar{\delta} =$ Perm $\{ \delta, X[i] \}$, where Perm is the permutation function.
8: $\alpha =$ Con $\{ \bar{\delta}, Y[i] \}$, where Con is the concatenation function.
9: $Cipher[i] = \alpha$
10: **end for**
11: Return the ciphertext $Cipher[c_1, c_2, c_3, \ldots, c_N]$, where c_i is the ciphertext of b_i

As shown in Algorithm 5, the working steps of the encryption process are as the following: First, we divide the bit sequence into 176-bits blocks and each block is subdivided into two 88-bits blocks. Then we calculate the number of 1's in each byte as well as the number of 1's for each 11 consecutive bits in the second half of the bit sequence. After that, we convert the plaintext into their corresponding binary codes and group them into blocks of 88-bits. This binary code is xored with the first block of the bit sequence since the additive cipher XOR is secure for same length of key-stream. Then, the permutation process is performed on each byte of xored plaintext. For example, if the number of 1's in the first byte of the second half of random bit sequence is 7, we mutate the 7th and 8th number bits in the first byte of the xored plaintext. The permuted plaintext is further concatenated as shown to generate the ciphertext. This concatenation operation is done repeatedly (e.g., $\{B_1\ B_2\ B_3\ B_4\}, \{B_2\ B_3\ B_4\ B_5\}, \ldots, \{B_{n-2}\ B_{n-1}\ B_n\ B_1\}$) so that each of the 11 bits performs the operation at least twice in order to add relatively fair diversity.

7.4.2 *Secure Data Aggregation*

In cluster-based WSN, a CH consumes more energy than a SN due to receiving, processing and retransmitting the data. This phase aims to avoid energy consumption of CH by allowing it to aggregate the encrypted data of their cluster members without having to decrypt them. As a result, the attacker won't be able to eavesdrop on the data from intermediate nodes, resulting in much stronger privacy than the traditional aggregation schemes. To do that, we use the addition property of the homomorphic encryption that allows arithmetic operations to be performed on ciphertext.

For example, Let $E()$ denotes encryption function. Let M is a group of messages under operation $\oplus$ and C is a group of messages under operation $\otimes$. $E()$ is a $(\oplus, \otimes)$ homomorphic encryption function if for $c_1 = E_{k_1}(m_1)$ and $c_2 = E_{k_2}(m_2)$, there exists a key k such that $c1 \otimes c_2 = E_k(m_1 \oplus m_2)$. In our scheme, each sensor node N_i senses data m_i, and encrypts it with k_i as shown in Eq. (7.7) and sends it to its CH. Where r is the round index in which the node produced the key k_i.

$$c_i = E(m_i, k_i^r, M) = m_i + k_i^r (mod M) \tag{7.7}$$

After receive the sensed data, the CH collects n messages and aggregates them by simply add them up as Eq. (7.8):

$$c = \sum_{i=1}^{|N|} c_i = \sum_{i=1}^{|N|} m_i + \sum_{i=1}^{|N|} k_i^r (mod M) \tag{7.8}$$

The last step after aggregating the data is to forward it to the BS. In order to arrange the aggregated data, CH will attach all nodes indexes at the end of the message. Thus, the last form of the transferred ciphertext C_T to BS will be as shown in Fig. 7.5 with total size $N * 176 + N * 13$ bits, where N is the count of nodes in the cluster.

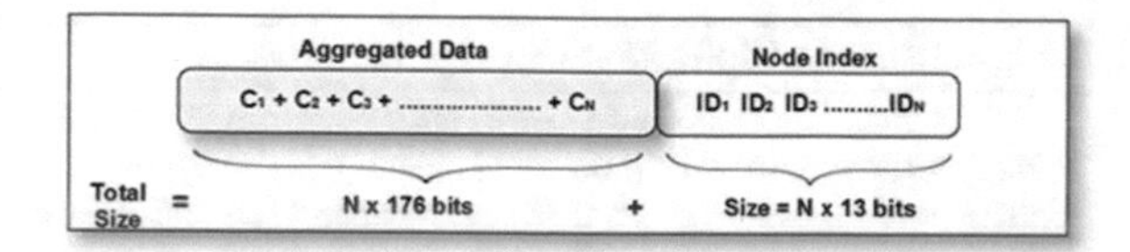

Fig. 7.5 The transfered ciphertext from CH to BS

7.4.3 Plaintext Retrieval Process

Data decryption process aims to get back the original data that has sent by every SN in the network. Thus, the BS has to do two main processes to get the plaintext: data deaggregation, and plaintext retrieval. As shown at Algorithm 5. there are two main purposes of data deaggregation step. First, the BS has to extract the IDs of the nodes away from the aggregated ciphertext C. Since the cluster information is stored at the BS, it knows the count of nodes inside the cluster and their IDs. As explained in secure data aggregation phase, the BS will start the separation process at index $N * 176$ which represents the end of the aggregated ciphertext. Second, the reverse process of Homomorphic Encryption Addition [39] will be used to get C_1, $C_2, \ldots$ and C_N.

As soon as the BS separates the ciphertext, the plaintext retrieval process will start. The BS will use the shared keys forward by CH to decrypt the data and get the plaintext. The decryption process will start by generating the cryptography key for each SN. The working steps of the decryption process that aims to extract the plaintext are as the following: Similarly to the encryption process, BS divides the encryption key into 176-bits blocks and each block is subdivided into two 88-bits blocks. Then it calculates the number of 1's in each byte x as well as the sum of 1's for each 11 consecutive bits y in the second half of the random bit sequence. After that, it converts the ciphertext into their corresponding binary codes and groups them into blocks of 88-bits. Then, the ciphertext will be concatenated with y to generate the permuted ciphertext. After that, the permutation process is performed on each byte of the ciphertext using x. For example, if the number of 1's in the first byte of the second half of bit sequence is 7, BS permutes the 7th and 8th number bits in the first byte of the ciphertext. Finally, the BS will get the plaintext by X-Nored the first 88-bits block of the bit sequence with the result of the permutation process.

7.5 Experimental Results and Discussion

The performance of the system was measured using the system throughput, network life time and the total energy consumption. In addition, we provide description about how our new schema prevents many kinds of attack to affect the network nodes.

Algorithm 6: Ciphertext Decryption Process

1: By receiving ciphertext C with size b bits from CH x, get the count of its cluster members N
2: η = the first ($N * 176$) bits of C, where η is the aggregated data.
3: $\mu = C - \eta$, where μ is the cluster members' IDs starting from bit index $(N * 176) + 1$ of C.
4: $\bar{\eta}$ = Deaggr (η), where Deaggr is the deaggregation function that revers the homomorphic operation.
5: C_i = Assign { $\bar{\eta}, \mu$ }, where C_i is the ciphertext of SN i
6: **for** $i = 1, 2, \ldots, N$ **do**
7: Partition C_i into 88-bits Q segments $S = \{s_1, s_2, \ldots, s_N\}$.
8: Generate the 176-bits bit sequence β
9: Divide β to two equal parts p_1 and p_2, 88-bits each.
10: Calculate the count of 1's for each byte $X[x_1, x_2, \ldots, x_8]$ as well as count of 1's for each 11 bits $Y[y_1, y_2, \ldots, y_8]$ in p_2
11: **for** $j = 1, 2, \ldots, Q$ **do**
12: α = Con { $\bar{\delta}, Y[i]$ }, where Con is the concatenation function.
13: $\delta = p_1 \otimes s_j$
14: $\bar{\delta}$ = Perm { $\delta, X[i]$ }, where Perm is the permutation function.
15: $Plain[j] = \bar{\delta}$
16: **end for**
17: Return the plaintext $Plain[m_1, m_2, m_3, \ldots, m_Q]$, where m_j is the plaintext of SN_j
18: **end for**

7.5.1 Experimental Settings

Our proposed method is simulated using the same characteristics of the sensor MICAz. The MICAz is based on the low-power 8-bit microcontroller ATmega128L with a clock frequency of 7.37 MHz and runs TinyOS as an event driven operating system. It also embeds a IEEE 802.15.4 compliant CC2420 transceiver with a claimed data rate of 250 kbps [79].

In this evaluation, the WSN has the following properties:

- There is one base station that receives data from nodes;
- Nodes are stationary and their positions are known;
- Provided with sufficient energy, each node can directly reach the base station;
- The characteristics and initial energy of each node are the same.

Table 7.1 lists the network parameters used in these experiments. In running GA to construct the network structure, we use the population size of 30 for 30 generations. The crossover probability and mutation probability are 0.8 and 0.006, respectively. The neighborhood distance δ is 20 m throughout these experiments when LSD is calculated.

The average performance of 10 repetitions is reported. In each experiment, nodes are randomly placed in the field and the base station is also randomly placed at a certain distance to the field center. Comparison studies are conducted with different state-of-the-art methods to evaluate both security and efficiency of the network.

Table 7.1 Network properties

Properties		Values
Number of nodes		100
Initial node energy		0.5 J
Idle state energy		50 nJ/bit
Data aggregation energy		5 nJ/bit
Amplification energy	$d \geq d_0$	10 pJ/bit/m^2
(cluster head to base station)	$d < d_0$	0.0013 pJ/bit/m^2
Amplification energy	$d \geq d_1$	$E_{fs}/10 = E_{fs1}$
(node to cluster head)	$d < d_1$	$E_{mp}/10 = E_{mp1}$

7.5.2 Memory Requirements

Operation speed determines time complexity and is a significant factor for performance evaluation. Based on [51], ATEMU is used to get the total CPU cycles required to encrypt 32 bytes data for MICA2 sensor mote. To evaluate the required memory size for each algorithm, TOSSIM has been used. Compared to state-of-the art methods, the results in the table indicate that our proposed algorithm performs better in terms of memory consumption (both RAM and ROM) as shown in Table 7.2. In case of CPU cycles, it can be seen that [51] is more efficient than the other protocols but still closed to our algorithm. Although our algorithm uses more time compared to [51], it is less than that of all other methods. Overall the proposed algorithm performed significantly better than other algorithms.

Table 7.2 Memory requirements and CPU cycles

Method	CPU (cycles)	Time (ms)	RAM (bytes)	ROM (bytes)
SkipJack	91224	12.353	292	7218
AES	68512	9.287	324	6994
LED	589652	78.972	378	5970
TWINE	128896	17.477	384	5280
BCC	91286	12.547	976	6240
Biswas	62396	8.547	542	5326
Our method	66201	8.619	281	3845

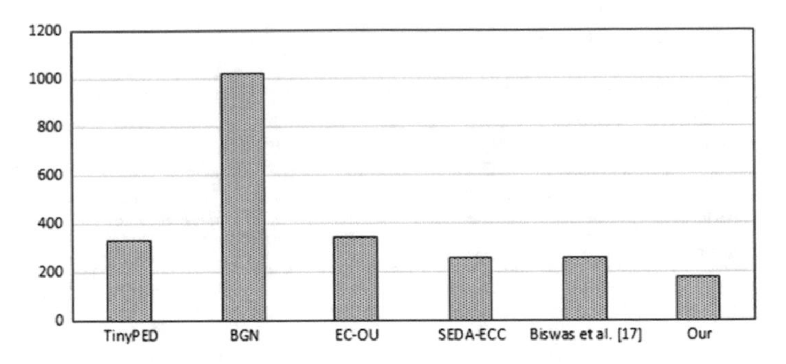

Fig. 7.6 Ciphertext size of different methods

7.5.3 Communication Overhead

The number of exchanged messages in each scheme is the same as each node needs to send two messages during the transmission process one HELLO message to establish the connection, and the other message for data transmission [39]. Thus, the communication overhead mainly depends on the ciphertext size of each algorithm if we assume that the number of message sending to the BS is the same. According to [39], the ciphertext size of SEDA-ECC is 256-bit. While EC-OU's ciphertext size is 341-bit, BGN's ciphertext size is 1,025-bit, and TinyPEDS's ciphertext size is 328-bit. In addition, the ciphertext size of the proposed algorithm in Biswas [51] is 256-bit. Our proposed method produces ciphertext with 176-bits size. Figure 7.6 shows the comparison of ciphertext sizes. So, we can conclude that, the communication overhead of our proposed schema is better than other schemes.

7.5.4 Energy Consumption and Network Lifetime

To evaluate the proposed schema from the energy consumption point of view, we compared it with SET-IBS, SL-LEACH, and SecLEACH methods as shown in Fig. 7.7. We find the proposed protocol has less power consumption and longest network lifetime. Compared with our method, the increase of power consumption in the state-of-the art methods started before 200 rounds with the loss of the first node as shown in Table 7.3. As a result, the other nodes faced more load due to the increase of power consumption which reduced the network life. On the other hand, the proposed protocol lost the first node at time. This reduced the power consumption and increased the network life time.

In order to evaluate the energy consumption, it is also interesting to examine the remaining energy of the sensor nodes. Ideally, we want to achieve the equal remaining energy in all nodes. Figure 7.8 illustrates the percentage of the remaining energy of sensor nodes at transmission rounds of 100 and 1000. Nevertheless, it is clear that the

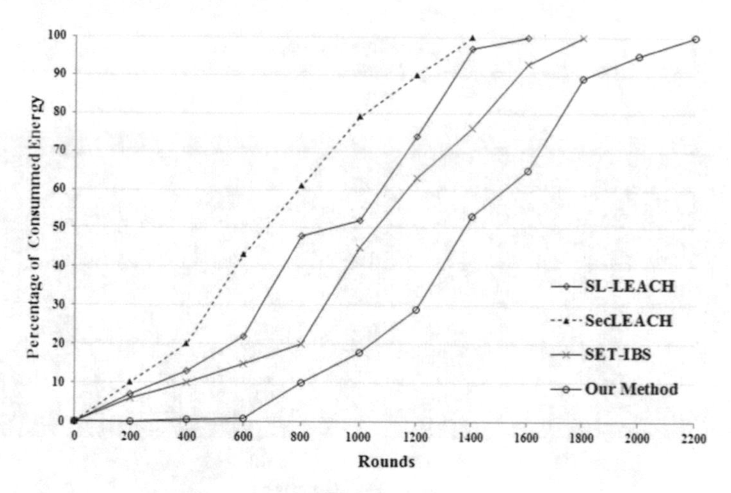

Fig. 7.7 Network energy consumption for different methods

Table 7.3 Network energy consumption. Each column gives the percentage of the consumed energy for all SENSOR NODES at a specific round

Round index	0	200	400	600	800	1000	1200	1400	1600	1800	2000	2200
SL-LEACH	0	7	13	22	48	52	74	97	100	100	100	100
SecLEACH	0	10	20	43	61	79	90	100	100	100	100	100
SET-IBS	0	6	10	15	20	45	63	76	93	100	100	100
Our method	0	0	0.5	1	10	18	29	53	65	89	95	100

remaining energy are fairly equal with some fluctuations. That is, as a consequence of our proposed method, the variance among remaining energy is quite low, which implies that the sensor nodes shared the burden of relaying messages and, hence, elongated the overall network life.

Figure 7.9 shows the comparison between our proposed method and the state-of-the art methods in terms of network life time. In addition, Table 7.4 presents the percentage of live sensor nodes throughout the life span of the WSN using different methods. The number of round is the average of 10 experiments with random sensor node placement. As shown, our proposed method yielded the largest number of rounds when the first sensor node dies. It is clear that our proposed method greatly extended the network life.

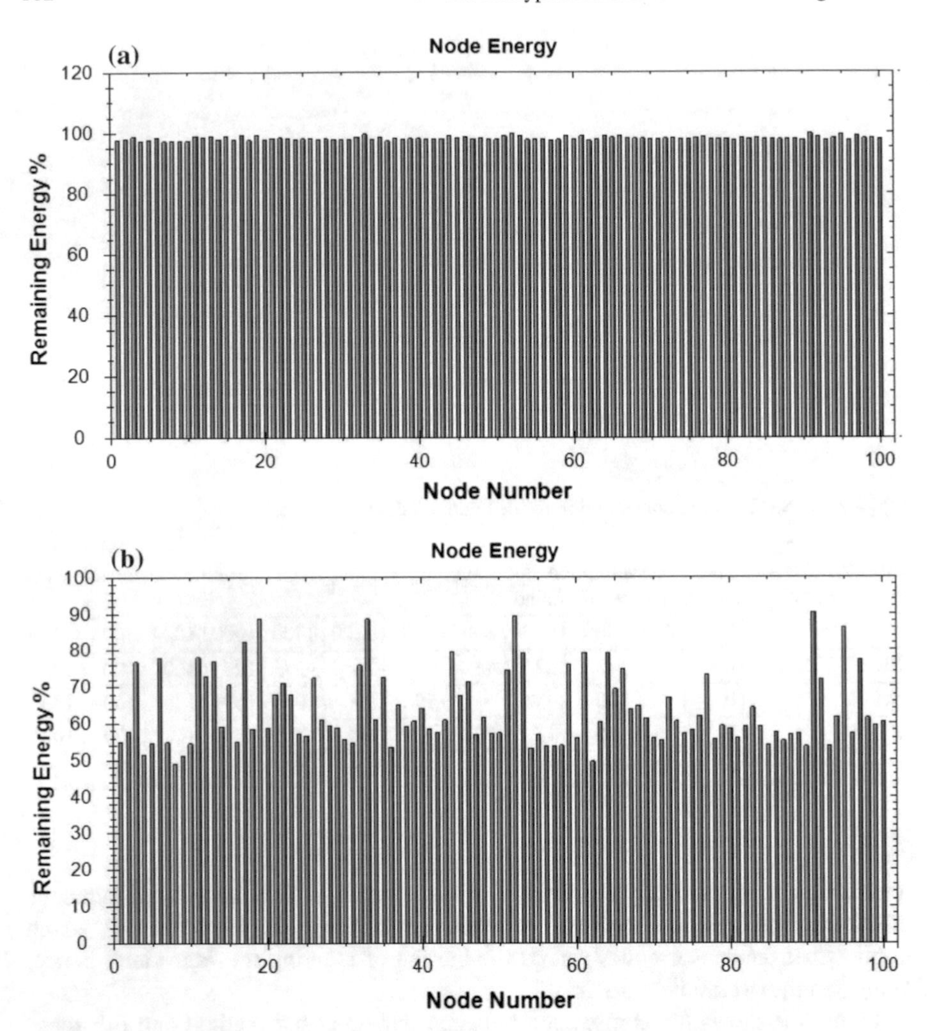

Fig. 7.8 The remaining energy of all sensor nodes **a** at network transmission round 100 and **b** at network transmission round 1000

Table 7.4 Network life span. Each column gives the number of network transmission rounds when the respective percentage of sensor nodes becomes unavailable due to depletion of energy

Percentage of alive nodes	100	90	80	70	60	50	40	30	20	10	0
SL-LEACH	509	592	698	780	893	1004	1082	1204	1290	1354	1402
SecLEACH	306	450	490	542	601	764	814	901	976	1200	1300
SET-IBS	591	695	801	894	1005	1100	1203	1360	1473	1600	1790
Our method	723	885	965	1060	1110	1202	1370	1530	17007	1830	2103
Our w/o encryption	1496	1620	1773	1832	1907	2100	2305	2501	2780	3400	5150

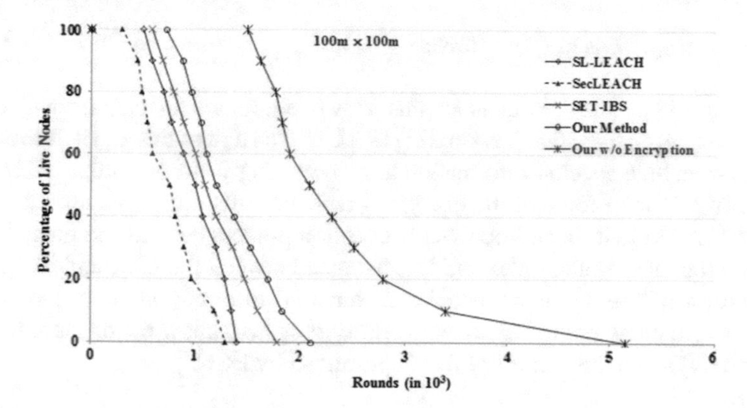

Fig. 7.9 Network lifetime figure that shows the percentage of the remaining number of live sensor nodes given the base station is placed at the corner of the field

7.5.5 *Security Analysis*

We have tested our proposed encryption scheme against various security attacks. Here, we describe some of the important security analysis results including Passive attack analysis, key-space analysis, and compromised CH attack analysis.

7.5.5.1 Passive Attacks

In the proposed protocol, the sensed data are encrypted by ECC and the homomorphic encryption algorithm which deals with eavesdropping. Thus, the passive adversaries cannot decrypt the eavesdropped message without the decryption key. Based on [37], properties of the proposed schema settle the countermeasures to passive attacks.

7.5.5.2 Key Space Analysis

Key space size is determined by the total number of different keys used in the encryption scheme. The key-space of a good encryption algorithm should be large enough to make brute-force attacks infeasible. In our proposed encryption scheme, the set of secret parameters is $\{ K_i, ID_i, D_{i,CH}, R_t \}$, where K_i is calculated based on $\{ x_i, y_i, A, B \}$ which are the parameters of the curve. K_i is represented in 128 bits and has the following range $K_i \in (1, 2^{128})$, $ID_i \in (1, 2^{13})$, $D_{i,CH} \in (1, 2^{15})$, and $R_t \in (1, 220)$. Therefore, the complete key-space of the proposed encryption scheme is $3.4e^{38}$. Hence, we conclude that brute force attack is not feasible for such a large key space.

7.5.5.3 Compromised CH Attacks

Compromised CH attack is an attack that tries to aggregate the data from all cluster nodes by deluding them that it is working as a CH. The aim of this attack is to analyze data and conclude specific information after receiving it. In our proposed schema, the role of CH is to forward the encrypted data from its cluster nodes and the BS without decrypting it. In addition, our encryption process depends on many factors, i.e., node location, round index and the distance between the node and its CH; this combination will requires any attacker to have all information about the network operations (to know round index), network topology (to know the distance between node and CH), and the secret key that is produced by ECC.

7.6 Conclusion

Forming network clusters is an effective way of improving scalability and longevity of WSN. However, security is a challenging issue in Cluster-based WSNs, since sensors are usually deployed with limited resources in unattended environments. Despite the great efforts in secure clustering of WSN, the dynamic nature of sensor network and numerous possible cluster configurations make searching for a secure and optimal network structure an open challenge. Due to its ability to provide high security with short key size, ECC is suitable for implementations on the devices with limited resources as sensor nodes in WSN. In this paper, we propose a novel encryption schema based on ECC and Homomorphic Encryption to secure data transmission in WSN with dynamic clustering nature. With the goal of optimizing the lifespan of the entire network, genetic algorithm is employed to search for the most suitable sensor nodes as the cluster heads to relay the messages to base station. Then, ECC is used to generate public and private keys for sensor nodes. The encryption key at each sensor node is 176-bits and is produced by combining the ECC key, identification number, distance to its CH, and the round index. To prevent the CH energy consumption as well as CH compromised attack, Homomorphic Encryption is used to allow CH to aggregate the encrypted data of its cluster members without having to decrypt them, to produce the final message that will be sent to the base station. Thus, it is prevent the attacker to know anything even if the CH is compromised, because CH is not responsible to encrypt messages. Compared with other methods, our experimental results demonstrated that our proposed method greatly improve the network performance in terms of lifetime, communication overhead, memory requirements, and energy consumption. In addition, it is prevent passive attack, CH compromised attack, and brute force attack.

References

1. Butun, I., Morgera, S. D., & Sankar, R. (2014). A survey of intrusion detection systems in wireless sensor networks. *IEEE Communications Surveys and Tutorials, 16*(1).
2. Padmavathi, D. G., & Shanmugapriya, M. (2009). A survey of attacks and security mechanisms and challenges in wireless sensor networks. *International Journal of Computer Science and Information Security, 4*(1).
3. Elhoseny, M., Yuan, X., Yu, Z., Mao, C., El-Minir, H. K., & Riad, A. M. (2015). Balancing energy consumption in heterogeneous wireless sensor networks using genetic algorithm. *IEEE Communications Letters, PP*(99), 1.
4. Hosseinabadi, A. A. R., Vahidi, J., Saemi, B., Sangaiah, A. K., & Elhoseny, M. (2018). Extended genetic algorithm for solving open-shop scheduling problem. *Soft Computing*. https://doi.org/10.1007/s00500-018-3177-y.
5. Elhoseny, M., Ramírez-González, G., Abu-Elnasr, O. M., Shawkat, S. A., Arunkumar, N., & Farouk, A. (2018). Secure medical data transmission model for IoT-based healthcare systems. *IEEE Access, PP*(99). https://doi.org/10.1109/ACCESS.2018.2817615.
6. Shehab, A., Elhoseny, M., Muhammad, K., Sangaiah, A. K., Yang, P., Huang, H., & Hou, G. (2018). Secure and robust fragile watermarking scheme for medical images. *IEEE Access*, 6(1), 10269–10278. https://doi.org/10.1109/ACCESS.2018.2799240.
7. Farouk, A., Batle, J., Elhoseny, M., Naseri, M., Lone, M., Fedorov, A., Alkhambashi, M., Ahmed, S. H., & Abdel-Aty, M., (2018). Robust general N user authentication scheme in a centralized quantum communication network via generalized GHZ states, *Frontiers of Physics, 13*, 130306. Springer. https://doi.org/10.1007/s11467-017-0717-3.
8. Elhoseny, M., Elkhateb, A., Sahlol, A., & Hassanien, A. E. (2018) Multimodal biometric personal identification and verification. In A. Hassanien, & D. Oliva (Eds.), *Advances in soft computing and machine learning in image processing*. Studies in Computational Intelligence, vol 730. Cham: Springer. https://doi.org/10.1007/978-3-319-63754-9_12.
9. Elhoseny, M., Essa, E., Elkhateb, A., Hassanien, A. E., & Hamad, A. (2018). Cascade multimodal biometric system using fingerprint and Iris patterns. In A. Hassanien, K. Shaalan, T. Gaber, & M. Tolba (Eds.), *Proceedings of the International Conference on Advanced Intelligent Systems and Informatics 2017, AISI 2017*. Advances in Intelligent Systems and Computing, Vol. 639. Cham: Springer. https://doi.org/10.1007/978-3-319-64861-3_55.
10. Elhoseny, M., Yuan, X., El-Minir, H. K., & Riad, A. M. (2016b). An energy efficient encryption method for secure dynamic WSN. *Security and Communication Networks, 9*(13), 2024–2031.
11. Elhoseny, M., Elminir, H., Riad, A., & Yuan, X. (2016a). A secure data routing schema for WSN using elliptic curve cryptography and homomorphic encryption. *Journal of King Saud University-Computer and Information Sciences, 28*(3), 262–275.
12. Elsayed, W., Elhoseny, M., Riad, A., & Hassanien, A. E. (2017). Autonomic self-healing approach to eliminate hardware faults in wireless sensor networks. In *International conference on advanced intelligent systems and informatics*, (pp. 151–160). Cham: Springer.
13. Elsayed, W., Elhoseny, M., Sabbeh, S., & Riad, A. (2017). Self-maintenance model for wireless sensor networks. *Computers and Electrical Engineering*. https://doi.org/10.1016/j.compeleceng.2017.12.022. (in Press).
14. Elhoseny, M., Yuan, X., El-Minir, H. K., & Riad, A. M. (2016). An energy efficient encryption method for secure dynamic WSN. *Security and Communication Networks, 9*(13), 2024–2031. https://doi.org/10.1002/sec.1459.
15. Elhoseny, M., Farouk, A., Zhou, N., Wang, M., Abdalla, S., & Batle, J. (2017a). Dynamic multi-hop clustering in a wireless sensor network: Performance improvement. *Wireless Personal Communications*, 1–21.
16. Elhoseny, M., Tharwat, A., Farouk, A., & Hassanien, A. E. (2017b). K-coverage model based on genetic algorithm to extend WSN lifetime. *IEEE Sensors Letters, 1*(4), 1–4.
17. Elhoseny, M., Tharwat, A., Yuan, X., & Hassanien, A. E. (2018). Optimizing K-coverage of mobile WSNs. *Expert Systems with Applications, 92*, 142–153. https://doi.org/10.1016/j.eswa.2017.09.008.

18. Elhoseny, M., Farouk, A., Batle, J., Shehab, A., & Hassanien, A. E. (2017). Secure image processing and transmission schema in cluster-based wireless sensor network. In *Handbook of research on machine learning innovations and trends*, (Chapter 45, pp. 1022–1040), IGI Global. https://doi.org/10.4018/978-1-5225-2229-4.ch045.
19. Elhoseny, Mohamed, Elleithy, Khaled, Elminir, Hamdi, Yuan, Xiaohui, & Riad, Alaa. (2015). Dynamic clustering of heterogeneous wireless sensor networks using a genetic algorithm towards balancing energy exhaustion. *International Journal of Scientific & Engineering Research*, *6*(8), 1243–1252.
20. Yuan, X., Elhoseny, M., El-Minir, H., & Riad, A. (2017). A genetic algorithm-based, dynamic clustering method towards improved WSN longevity. *Journal of Network and Systems Management*, *25*(1), 21–46.
21. Guo, W., & Zhang, W. (2014). A survey on intelligent routing protocols in wireless sensor networks. *Journal of Network and Computer Applications*, *38*, 185–201.
22. Ahmed, G., Khan, N. M., & Ramer, R. (2008). Cluster head selection using evolutionary computing in wireless sensor networks. In *Progress in electromagnetics research symposium*, (pp. 883–886).
23. Bhaskar, N., Subhabrata, B., & Soumen, P. (2010). Genetic algorithm based optimization of clustering in ad-hoc networks. *International Journal of Computer Science and Information Security*, *7*(1), 165–169.
24. Asim, M., & Mathur, V. (2013). Genetic algorithm based dynamic approach for routing protocols in mobile ad hoc networks. *Journal of Academia and Industrial Research*, *2*(7), 437–441.
25. Karimi, A., Abedini, S., Zarafshan, F., & Al-Haddad, S. (2013). Cluster head selection using fuzzy logic and chaotic based genetic algorithm in wireless sensor network. *Journal of Basic and Applied Scientific Research*, *3*(4), 694–703.
26. Elhoseny, M., Tharwat, A., & Hassanien, A. E. (2017c). Bezier curve based path planning in a dynamic field using modified genetic algorithm. *Journal of Computational Science*. https://doi.org/10.1016/j.jocs.2017.08.004.
27. Metawa, N., Hassan, M. K., & Elhoseny, M. (2017). Genetic algorithm based model for optimizing bank lending decisions. *Expert Systems with Applications*, *80*, 7582. https://doi.org/10.1016/j.eswa.2017.03.021.
28. Elhoseny, M., Shehab, A., & Yuan, X. (2017). Optimizing robot path in dynamic environments using genetic algorithm and bezier curve. *Journal of Intelligent & Fuzzy Systems*, *33*(4), 2305–2316. IOS-Press. https://doi.org/10.3233/JIFS-17348.
29. Rana, K., & Zaveri, M. (2013). Synthesized cluster head selection and routing for two tier wireless sensor network. *Journal of Computer Networks and Communications*, *13*(3).
30. Bayrakl, S., & Erdogan, S. (2012). Genetic algorithm based energy efficient clusters in wireless sensor networks. *Procedia Computer Science*, *10*, 247–254.
31. Attea, B. A., & Khalil, E. A. (2012). A new evolutionary based routing protocol for clustered heterogeneous wireless sensor networks. *Applied Soft Computing*, *12*(7), 1950–1957.
32. Wu, Y., & Liu, W. (2013). Routing protocol based on genetic algorithm for energy harvesting-wireless sensor networks. *IET Wireless Sensor Systems*, *3*(2), 112–118.
33. Hussain, S., Matin, A. W., & Islam, O. (2007). Genetic algorithm for energy efficient clusters in wireless sensor networks. In *International conference on information technology*.
34. Lotf, J. J., Hossein, S., & Ghazan, N. (2011). Overview on wireless sensor networks. *Journal of Basic and Applied Scientific Research*, *11*(1), 2811–2816.
35. Lalitha, T., & Umarani, R. (2012). Energy efficient cluster based key management technique for wireless sensor network. *International Journal of Advances in Engineering and Technology*, *3*(2), 186–190.
36. Ganesh, S., & Amutha, R. (2013). Efficient and secure routing protocol for wireless sensor networks through snr based dynamic clustering mechanisms. *Journal of Communications and Networks*, *15*(4), 422–429.
37. Huang, L., Jie, L., & Guizani, M. (2014). Secure and efficient data transmission for cluster-based wireless sensor networks. *IEEE Transactions on Parallel and Distributed Systems*, *25*(3), 750–761.

38. Xiao, W., Li, Y., & Ke, C. (2005). SLEACH secure lowenergy adaptive clustering hierarchy protocol for wireless sensor networks. *Wuhan University Journal of Natural Sciences, 10*(1), 127–131.
39. Zhou, Q., Yang, G., & He, L. (2014). A secure enhanced data aggregation based on ECC in wireless sensor network. *Sensors Journal, 14*(4), 6701–6721.
40. Girao, J., Westhoff, D., Mykletun, E., & Araki, T. (2007). TinyPEDS tiny persistent encrypted data storage in asynchronous wireless sensor networks. *Ad Hoc Networks, 5*(7), 1073–1089.
41. Oliveira, L., Ferreira, A., Vilaca, M., Wong, H., Bern, M., Dahab, R., et al. (2007). Secleach-on the security of clustered sensor networks. *Signal Processing, 87*(12), 2882–2895.
42. Boneh, D., Goh, E. J., & Nissim, K. (2005). Evaluating 2-DNF formulas on ciphertexts. In *The 2nd international conference on theory of cryptography*, (pp. 325–341).
43. Mykletun, E., Girao, J., & Westhoff, D. (2006). Public key based cryptoschemes for data concealment in wireless sensor networks. In *The IEEE international conference on communications*, (pp. 2288–2295). Istanbul: IEEE.
44. Bogdanov, A., Khovratovich, D., & Rechberger, C. (2011). Biclique cryptanalysis of the full AES. *Advances in Cryptology, 7073*, 344–371.
45. Xiao-Jun, T., Zhu, W., & Ke, Z. (2012). A novel block encryption scheme based on chaos and an Sbox for WSNs. *Journal of Chinese Physics, 21*(2).
46. Eryümaz, E., Erturk, I., & Atmaca, S. (2009). Implementation of Skipjack cryptology algorithm for WSNs using FPGA. In *International conference on application of information and communication technologies*, (pp. 1–5). IEEE.
47. Biham, E., Biryukov, A., & Shamir, A. (2005). Cryptanalysis of Skipjack reduced to 31 rounds using impossible differentials. *Journal of Cryptology, 18*(4), 291–311.
48. Suzaki, T., Minematsu, K., Morioka, S., & Kobayashi, E. (2013). TWINE: a lightweight block cipher for multiple platforms. *Selected Areas in Cryptography, 7707*, 339–354.
49. Karakoca, F., Demircia, H., & Harmancib, A. (2013). Biclique cryptanalysis of lblock and twine. *Information Processing Letters, 113*(12), 423–429.
50. Isobe, T., & Shibutani, K. (2012). Security analysis of the lightweight block ciphers XTEA and LED and Piccolo. *Information Security and Privacy, 7372*, 71–86.
51. Biswas, K., Muthukkumarasamy, V., & Singh, K. (2015). An encryption scheme using chaotic map and genetic operations for wireless sensor networks. *IEEE Sensors Journal, PP*(99), 1.
52. Liu, Y., & Tian, S. (2012). Design and statistical analysis of a new chaos block cipher for WSN. *Communications in Nonlinear Science and Numerical Simulation, 17*(8), 3267–3278.
53. Saminathan, A., & Karthik, S. (2013). Development of an energyefficient and secure and reliable wireless sensor networks routing protocol based on data aggregation and user authentication. *American Journal of Applied Sciences, 10*(8), 832–843.
54. Singh, M., & Hussain, M. (2010). A top-down hierarchical multi-hop secure routing protocol for wireless sensor networks. *International Journal of Ad hoc and Sensor and Ubiquitous Computing, 1*(2).
55. Elhoseny, M., Yuan, X., ElMinir, H., & Riad, A. (2014). Extending self-organizing network availability using genetic algorithm. In *ICCCNT*. China: Hefei.
56. Tharwat, A., Mahdi, H., Elhoseny, M., & Hassanien, A. E. (2018). Recognizing human activity in mobile crowdsensing environment using optimized k-NN algorithm. *Expert Systems With Applications*. https://doi.org/10.1016/j.eswa.2018.04.017. Accessed 12 April 2018.
57. Tharwat, A., Elhoseny, M., Hassanien, A. E., Gabel, T., & Kumar, A. (2018). Intelligent Bezir curve-based path planning model using chaotic particle swarm optimization algorithm. *Cluster Computing*, (pp. 1–22). Springer. https://doi.org/10.1007/s10586-018-2360-3.
58. Rizk-Allah, R. M., Hassanien, A. E., & Elhoseny, M. (2018). A multi-objective transportation model under neutrosophic environment. *Computers and Electrical Engineering*. Elsevier. https://doi.org/10.1016/j.compeleceng.2018.02.024. (in Press).
59. Batle, J., Naseri, M., Ghoranneviss, M., Farouk, A., Alkhambashi, M., & Elhoseny, M. (2017). Shareability of correlations in multiqubit states: Optimization of nonlocal monogamy inequalities. *Physical Review A, 95*(3), 032123. https://doi.org/10.1103/PhysRevA.95.032123.

60. El Aziz, M. A., Hemdan, A. M., Ewees, A. A., Elhoseny, M., Shehab, A., Hassanien, A. E., & Xiong, S. (2017). Prediction of biochar yield using adaptive neuro-fuzzy inference system with particle swarm optimization. In *2017 IEEE PES PowerAfrica conference*, (pp. 115–120), June 27–30, 2017. Accra-Ghana: IEEE. https://doi.org/10.1109/PowerAfrica.2017.7991209.
61. Ewees, A. A., El Aziz, M. A., & Elhoseny, M. (2017). Social-spider optimization algorithm for improving ANFIS to predict biochar yield. In *8th International conference on computing, communication and networking technologies (8ICCCNT)*, July 3–5, 2017. Delhi-India: IEEE.
62. Metawa, N., Elhoseny, M., Hassan, M. K., & Hassanien, A. E. (2016). Loan portfolio optimization using genetic algorithm: A case of credit constraints. In *Proceedings of 12th international computer engineering conference (ICENCO)*, (pp. 59–64). IEEE.https://doi.org/10.1109/ICENCO.2016.7856446.
63. Shehab, A., Elhoseny, M., El Aziz, M. A., & Hassanien, A. E. (2018). Efficient schemes for playout latency reduction in P2P-VoD systems. In A. Hassanien, & D. Oliva (Eds.), *Advances in soft computing and machine learning in image processing*. Studies in Computational Intelligence, Vol. 730. Cham: Springer. https://doi.org/10.1007/978-3-319-63754-9_22.
64. Elhoseny M., Nabil A., Hassanien A. E., & Oliva, D. (2018). Hybrid rough neural network model for signature recognition. In A. Hassanien, & D. Oliva (Eds.), *Advances in soft computing and machine learning in image processing*. Studies in Computational Intelligence, Vol. 730. Cham: Springer. https://doi.org/10.1007/978-3-319-63754-9_14.
65. Abdeldaim, A. M., Sahlol, A. T., Elhoseny, M., & Hassanien, A. E. (2018). Computer-aided acute lymphoblastic Leukemia diagnosis system based on image analysis. In A. Hassanien, & D. Oliva (Eds.), *Advances in soft computing and machine learning in image processing*. Studies in Computational Intelligence, Vol. 730. Cham: Springer. https://doi.org/10.1007/978-3-319-63754-9.
66. Elhoseny, H., Elhoseny, M., Riad, A. M., & Hassanien, A. E. (2018). A framework for big data analysis in smart cities. In A. Hassanien, M. Tolba, M. Elhoseny, & M. Mostafa (Eds.), *AMLTA 2018 the international conference on advanced machine learning technologies and applications (AMLTA2018)*. Advances in Intelligent Systems and Computing, Vol. 723. Cham: Springer. https://doi.org/10.1007/978-3-319-74690-6_40.
67. Elhoseny, M., Shehab, A., & Osman, L. (2018). An empirical analysis of user behavior for P2P IPTV workloads. In A. Hassanien, M. Tolba, M. Elhoseny, & Mostafa M. (Eds.), *AMLTA 2018 the international conference on advanced machine learning technologies and applications (AMLTA2018)*. Advances in Intelligent Systems and Computing, Vol. 723. Cham: Springer. https://doi.org/10.1007/978-3-319-74690-6_25.
68. Wang, M. M., Qu, Z. G., & Elhoseny, M. (2017). Quantum secret sharing in noisy environment. In X. Sun, H. C. Chao, X. You, & E. Bertino (Eds.), *Cloud computing and security, ICCCS 2017*. Lecture Notes in Computer Science, Vol. 10603. Cham: Springer. https://doi.org/10.1007/978-3-319-68542-7_9.
69. Elsayed, W., Elhoseny, M., Riad, A. M., & Hassanien, A. E. (2018). Autonomic self-healing approach to eliminate hardware faults in wireless sensor networks. In A. Hassanien, K. Shaalan, T. Gaber, & M. Tolba (Eds.), *Proceedings of the international conference on advanced intelligent systems and informatics 2017, AISI 2017*. Advances in Intelligent Systems and Computing, Vol. 639. Cham: Springer. https://doi.org/10.1007/978-3-319-64861-3_14.
70. Abdelaziz, A., Elhoseny, M., Salama, A. S., Riad, A. M., & Hassanien, A. E. (2018). Intelligent algorithms for optimal selection of virtual machine in cloud environment, towards enhance healthcare services. In A. Hassanien, K. Shaalan, T. Gaber, & M. Tolba (Eds.), *Proceedings of the international conference on advanced intelligent systems and informatics 2017, AISI 2017*. Advances in Intelligent Systems and Computing, Vol. 639. Cham: Springer. https://doi.org/10.1007/978-3-319-64861-3_27.
71. Shehab A., Ismail A., Osman L., Elhoseny M., & El-Henawy I. M. (2018). Quantified self using IoT wearable devices. In A. Hassanien, K. Shaalan, T. Gaber, & M. Tolba (Eds.), *Proceedings of the International Conference on Advanced Intelligent Systems and Informatics 2017, AISI 2017*. Advances in Intelligent Systems and Computing, Vol. 639. Cham: Springer. https://doi.org/10.1007/978-3-319-64861-3_77.

72. Elhoseny, M., Abdelaziz, A., Salama, A. S., Riad, A. M., Muhammad, K., & Sangaiah, A. K. (2018). A hybrid model of internet of things and cloud computing to manage big data in health services applications. *Future Generation Computer Systems*. Elsevier. (in Press).
73. Abdelaziz, A., Elhoseny, M., Salama, A. S., & Riad, A. M. (2018). A machine learning model for improving healthcare services on cloud computing environment. *Measurement*, *119*, 117–128. https://doi.org/10.1016/j.measurement.2018.01.022.
74. Darwish, A., Hassanien, A. E., Elhoseny, M., Sangaiah, A. K., & Muhammad, K. (2017). The impact of the hybrid platform of internet of things and cloud computing on healthcare systems: Opportunities, challenges, and open problems. *Journal of Ambient Intelligence and Humanized Computing*. Springer. https://doi.org/10.1007/s12652-017-0659-1.
75. Yuan, X., Li, D., Mohapatra, D., & Elhoseny, M. (2017). Automatic removal of complex shadows from indoor videos using transfer learning and dynamic thresholding. *Computers and Electrical Engineering*. https://doi.org/10.1016/j.compeleceng.2017.12.026. (in Press).
76. Sajjad, M., Nasir, M., Muhammad, K., Khan, S., Jan, Z., Sangaiah, A. K., Elhoseny, M., & Baik, S. W. (2017). Raspberry Pi assisted face recognition framework for enhanced law-enforcement services in smart cities. *Future Generation Computer Systems*. Elsevier. https://doi.org/10.1016/j.future.2017.11.013.
77. Elhoseny, M., Hosny, A., Hassanien, A. E., Muhammad, K., & Sangaiah, A. K. (2017). Secure automated forensic investigation for sustainable critical infrastructures compliant with green computing requirements. *IEEE Transactions on Sustainable Computing*, *PP*(99). https://doi.org/10.1109/TSUSC.2017.2782737.
78. Houssain, H., Badra, M., & Somani, T. (2012). Software implementations of elliptic curve cryptography in wireless sensor networks. *Journal of Communication and Computer*, *9*, 712–720.
79. De Meulenaer, G., Gosset, F., Standaert, F. X., & Pereira, O. (2008). On the energy cost of communication and cryptography in wireless sensor networks. In *IEEE international conference on wireless and mobile computing and networking and communications*, (pp. 580–585). IEEE.

Chapter 8
Using Wireless Sensor to Acquire Live Data on a SCADA System, Towards Monitoring File Integrity

Abstract SCADA systems are network presence systems that face significant threats and attacks. After an attack occurred, SCADA requires forensic investigation to understand the cause and effects of the intrusion or disruption on the systems services. However, forensic investigators cannot turn it off during acquiring the live data that is required for the investigation and analysis process. That is because the systems services need to be continuously operational. Despite the great efforts to acquire live data on SCADA systems, the continuously change of this type of data and the risk on the systems services make it a big challenge. In this proposal, we suggest a new method to acquire live data on a SCADA system using wireless sensor network. The proposed idea attempts to monitor file integrity and acquire live data in a way that minimizes risk to the systems services. In addition, it aims to help Forensic investigators by guarantee early data acquisition after incident and digital evidence validity as well.

8.1 Introduction

Supervisory control and data acquisition (SCADA) system is a computer system spread over a wide area in order to remotely control and monitoring a specific process in a working field [1, 2]. Recently, SCADA is used in different smart applications [3–10] which can be can be industrial, infrastructure or facility-based. Currently, the challenges of the SCADA systems are related to quality attributes (or non-functional attributes) [11]. The extent to which the system possesses a desired combination of quality attributes such as usability, performance, reliability, and security indicates the success of the design and the overall quality of the software system [12–16]. Software systems quality attributes are the main challenges for developers and designers [11]. One of these challenges is the live data acquisition which involves acquiring both volatile and non-volatile data. Of course, the challenge is bigger in case of acquiring the volatile data, i.e., the contents of physical memory, without affecting the system performance.

Since SCADA systems are network presence systems and face significant threats and vulnerabilities, many algorithms that try to maximize the security procedures

M. Elhoseny and A. E. Hassanien, *Dynamic Wireless Sensor Networks*, Studies in Systems, Decision and Control 165, https://doi.org/10.1007/978-3-319-92807-4_8

of a SCADA system have been developed. But little is available that deals specifically with live data acquisition for digital forensic. Techniques for forensic analysis were adapted and tested on live SCADA, resulting in recommendations for successful detection and recovery after an incident. With adequate preparation and the appropriate response planning and execution, it is possible to successfully perform a forensic analysis for a SCADA compromise. In [17], a SCADA Forensics Architecture is proposed for applying IT forensic techniques to SCADA servers. However, many applications require a the live data acquisition represents the main challenge of this architecture.

In [18], recommendations for data capture and retention are provided. These recommendations are built based on the logging capabilities of a typical SCADA architecture and the analytical techniques and investigative tools that may help develop forensic readiness to the level of the current threat environment requirements. In [19], a new software tool for real time data acquisition is proposed. This method aims to detect, prevent, or evaluate critical situations. However, the implementation details of this tool are unclear. In addition, the author mentioned that the proposed tool affects the performance of the SCADA system by consuming its resources, during the data acquisition process. Accordingly, an optimization algorithm [20, 22–27] is needed to balance between the accuracy of live data collection and the efficiency of the SCADA system performance.

Therefore, we propose a new method for live data acquisition using Wireless sensor network (WSN) to acquire live data without affecting the SCADA systems performance. The reason why is because WSN has been widely used in several applications [28–30] such as health care, transportation management, military surveillance, etc. by placing battery operated sensors in open fields without human attendance, and acquire data contiguously over time [31–36].

The expected contribution of the proposed method is twofold: first, a WSN-based secure live data acquiring [37–41] method for SCADA systems. WSN collects the acquired data and transmits it to a separate server in order to process it. This process minimizes the risk on the systems services and helps Forensic investigators to conduct their work by guarantee early data acquisition after incident, and digital evidence validity as well. Second, the proposed system represents a second line of data security in the SCADA environment. Moreover, there is no need to change the security procedures applied at SCADA due to its ability to work with any existing SCADA system regardless its architecture.

8.2 Research Objectives

The proposed model concerns three important requirements of modern SCADA applications: Adaptivity, Security, and Forensic Investigation, combined in one framework. The expected contributions of this research are:

- **Adaptivity**: a new approach for dynamically reorganizing (re-structuring) the complex real-life SCADA application containing a huge amount of data to adapt environment changes. Adaptivity will be realized using self-organizing WSN.
- **Security**: a new trust-based soft security approach specially designed for large-scale cyberspace applications that developed based on WSN. Security will be realized using a novel trust-based security mechanism for self-organizing WSN.
- **Forensics**: a new approach for feasible and effective forensic investigation in an automated manner based on intelligent life data collection and self-organizing WSN, which can be self-organized by an optimization algorithm such as Genetic Algorithm [42]. Forensic investigation will be realized based on intelligent agents and wireless sensors.

It is intended to make the proposed triple fold framework as a general-purpose framework that can be used for security and forensically engineering most of modern computing paradigms (i.e., CPS, STS, SoS, CAS, etc.). However, the main concerned systems here are CPS such as SCADA. The above three contribution approaches combined in one framework (as demonstrated in Fig. 8.1). The first contribution is tailored to the adaptivity of a multi-agent SCADA system, in contrast to initial Multiagent systems research, which concerned individual agents aspects such as agents architectures, agents mental capabilities, behaviors, etc., the current research trend of Multiagent System (MAS) is actively interested in the adaptivity, environment, openness and the dynamics of these systems. Also, there is a great attention towards the MAS technique as a way to design self-organized systems. In open environments, agents must be able to adapt towards the most appropriate organizations according to the environment conditions and their unpredictable changes. MAS that have the

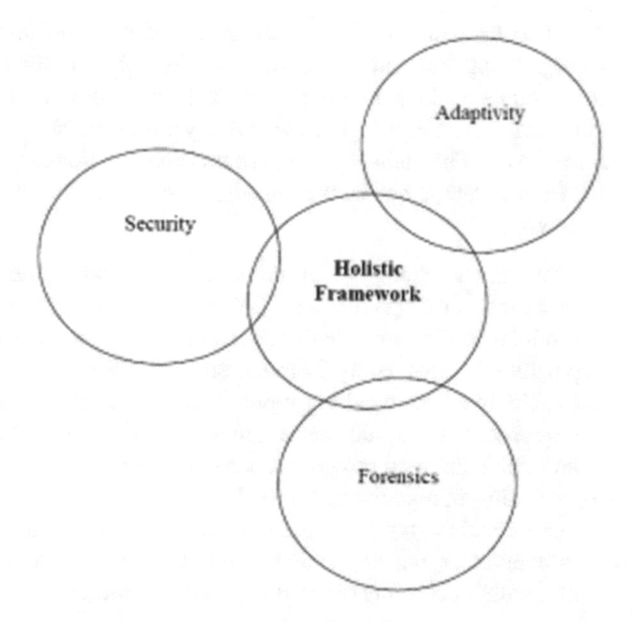

Fig. 8.1 The proposed holistic framework

ability to dynamically reorganize (regardless of the type of reorganization, self or enforced) will be adaptive enough to survive against their dynamic and continuously changing working environments. Dynamic reorganization can take many forms, for instance, agents can dynamically change their roles, behaviors, locations, acquaintance networks, or the whole system organization can be dynamically changed. Organizational models have been recently used in agent theory for modeling coordination in open systems and to ensure social order in MAS applications. They are used as a modeling technique to represent the social organization of MAS.

The MAS organizational models aim to enable the agents to dynamically reorganize to adapt environment dynamic and unpredictable changes. They give attention not only to the system micro level (individual agents) but also to the system macro level (the global system structure and behavior). In this research project, it is intended to use a recent MAS organizational model called NOSHAPE presented in [43, 44] for dynamically reorganizing MAS to adapt the environment dynamic changes and for agents localization. The NOSHAPE model was tested and applied for the development of adaptive large-scale agent-based SCADA system [45]. The second contribution approach is tailored towards cyberspace security; it proposes a new trust-based technique for securing agent-based Internet applications such as CPS. The third contribution approach is tailored towards the forensic investigation of modern Internet applications based on intelligent agents and WSN. In the next subsections, we present the project contributions in some details.

8.2.1 The First Contribution

The first contribution is the adoption of the adequate approach for dynamically re-organizing (re-structuring) the complex agent-based real-life application which might contain a huge number of agents to adapt environment changes. The proposed approach will be based upon a previously presented MAS organizational model called NOSHAPE. The main objectives of this contribution are targeted towards adaptivity, modularity, agent localization, agents interactions coordination as summarized in the following:

- Providing a decomposition model for partitioning the system with large-number of agents into organizations (or groups) of agents so that the system becomes modular in the sense that the system administrator becomes able to modify any partition independently from the other partitions.
- Providing a matchmaking model that enables the agents to find each other as in complex systems with large number of agents it is difficult for an agent to find another agent who provides a particular service.
- Providing an organizational model for enabling dynamic reorganization of system agents against unpredictable working environment dynamic changes. The system dynamic reorganization can be realized by adding or removing agents as required and without shutting down the system (at will).

- Providing a coordination model for aligning agents behaviors and interactions. In other words, the model enables system agent to decide why and when to interact and do actions.

The implementation challenge of the first contribution is related to the used agent platform, which provides the execution environment for agents. JADE [46] is a software framework fully implemented in Java language. It simplifies the implementation of multi-agent systems through a middleware that claims to comply with FIPA specifications [47] and through a set of tools that supports the debugging and deployment phase. JADE as a middle-ware framework provides all the required low-level services to enable flexible agents interactions. Each JADE platform can be distributed across many machines. Further, each JADE platform contains by default two important services, the agent management services (AMS), which provides the white page service for the platform agents and concerns basic issues such as agents creation, termination, identification, and so on. The other service is the yellow page service provided by a system agent called the Directory Facilitator (DF), which enables agents to register/deregister their services so that other agents can find them by contacting the DF agent. The JADE agent platform can be distributed across machines with different operating systems and the configuration can be controlled via a remote GUI. The configuration can be even changed at run-time by creating new agents and moving agents from one machine to another one when required. JADE is distributed in open source. Also, the need to permit the deployment of agents in low cost resource-constrained devices led to the creation of the Lightweight Extensive Agent Platform (LEAP) add-on [48] for the JADE platform. This lighter version opens the door to the participation of devices with more limited computational resources, and can be exploited in smart phones and microcontrollers [49].

8.2.2 The Second Contribution

The second contribution is proposing a novel trust-based soft security approach specially designed for large-scale Internet-based applications developed using MAS. As the Internet is supposed to be used as a communication medium for agents interactions, the security will become an important challenge. There are many security approaches and methods proposed and implemented globally in order to secure CPS, along with areas such as social engineering, security standards, vendor control, as well as access control implementation, etc. However, in addition to these areas, another important concept, namely trust, is significant in ensuring secure and reliable communications in CPS [50]. Trust-based security techniques are designed to ensure that the security and reliability of the network will not be compromised after a particular user is permitted to join the network. In [51], trust is defined as a psychological state comprising the intention to accept vulnerability based upon positive expectations of the intentions or behavior of another. In this research project, it is intended to design a new trust-based soft security model for securing agents interactions. Soft security is concerned with human-like attitudes such as trust and reputation, this approach is very suitable to be used with agents which behave similar to humans.

The objective of this research project contribution is to design a new trust model to be used for securing agents interactions in the proposed framework so that only trusted agents could interact with system agents. The proposed trust-based authentication technique is realized not only in the individual agents level but also in the agent organizations level. An agent organization will only accept the dynamic overlap requests that come from trusted organizations. Trusted agent organizations are the organizations with trust value above a certain threshold (careful trust) values (i.e. 0.6) predefined initially by the system administrator but then they change according to the agent organization experience and according to the context. That is considered as a double protection because before the individual agents could securely interact their host organization must securely overlap.

8.2.3 The Third Contribution

The third contribution is proposing a novel approach for feasible and effective forensic investigation in an automated manner based on intelligent agents and self-organizing wireless sensor networks. The main objectives of this contribution are:

- Virtualization of the forensic investigation process to solve the limitations of the traditional fully human-based forensic investigation approaches. Virtualization is the process of creating a software-based (or virtual) representation of something rather than a physical one, so we hope to replace (not completely!) the human forensic investigator with a software agent that resides on a personal computer or a personal digital assistant (PDA). Actually, the software agent will not replace the human investigator completely but partially. In other words, a large percentage of the work and activity of the CPS human investigator will be done by the virtual software agent. This approach will solve many challenges that facing the human investigator such inspecting a large number of CPS components with high geographical distribution. In large CPS the human investigator might do that in a month but the virtual investigator (the software agent) can do that in one hour depending on the communication network speed.
- Making CPS under continuous forensic supervision in addition to the operational supervision. The human investigator will only appear in case of a catastrophic incident, but in normal conditions he will be in bed! In contrary, the virtual investigator will run continuously (it is assumed!) without stopping and without sleeping, and that will be very useful in catastrophic incidents because the virtual investigator has a very useful information about the incident.

- Enabling any-time any-where forensic access to CPS. The proposed framework will provide the human investigator the possibility to access the CPS network from his home, office, street, car, etc., with no necessity to be onsite. This is a very useful attribute because it saves a big amount of critical time in catastrophic situations.

8.3 Background Information

In the past, real-life applications (industrial or commercial) were designed with the monolithic (one piece) architecture in which functionally distinguishable aspects (for example data input and output, data processing, error handling, and the user interface) are all interwoven, rather than containing architecturally separate components (Fig. 8.2a). Although, the monolithic architecture has some advantages such as improved performance and efficiency, single deployment unit, no communication overhead, but from the other hand it has many disadvantages such as low scalability, low maintainability, low flexibility, low fault tolerance, etc. These disadvantages and the increase in the complexity of the application functionalities had led to the transfer to the distributed architecture (Fig. 8.2b) in which the application functionalities

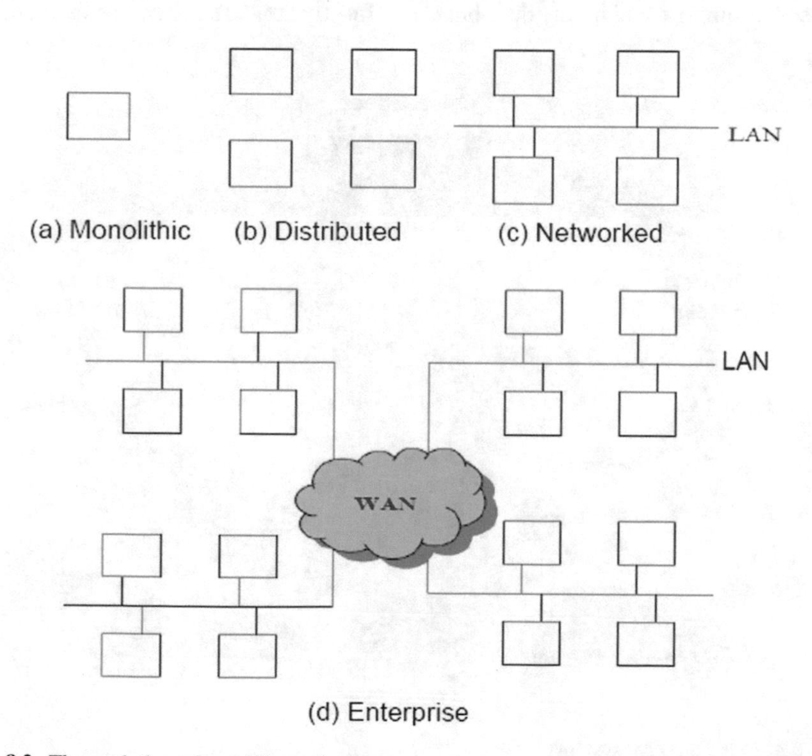

Fig. 8.2 The evolution of real-life applications

were separated and distributed into many components. In distributed applications, information processing is distributed over several components rather than confined to a single machine. Then for coordination purposes, the applications distributed components were connected through Local Area Networks (LAN), as demonstrated in Fig. 8.2c. The network architecture refers to the layout of the network, consisting of the hardware, software, connectivity, communication protocols and mode of transmission (wired or wireless). The types of networks are classified (according to their covered range) into: LAN (Local Area Network), MAN (Medium Area Networks) and WAN (Wide Area Network). WAN (Fig. 8.2d) are a telecommunication network or computer network that extends over a large geographical distance. WAN are used to connect LAN and other types of networks together, so that users and computers in one location can communicate with users and computers in other locations. Many WAN are built for one particular organization and are private. Others, built by Internet service providers, provide connections from an organization's LAN to the Internet. The Internet is the universal enterprise WAN that is currently getting great attention as a universal communication medium for connecting most of modern real-life activities.

The emerging of new engineering research areas such as CPS, STS, SoS, CAS, Unicom, AC, IoT, and IoE are a clear evidence of the emerging new demands and requirements of modern real-life applications. Mostly, these systems are open, heterogeneous, complex, and highly distributed and mostly they use Internet as their under-

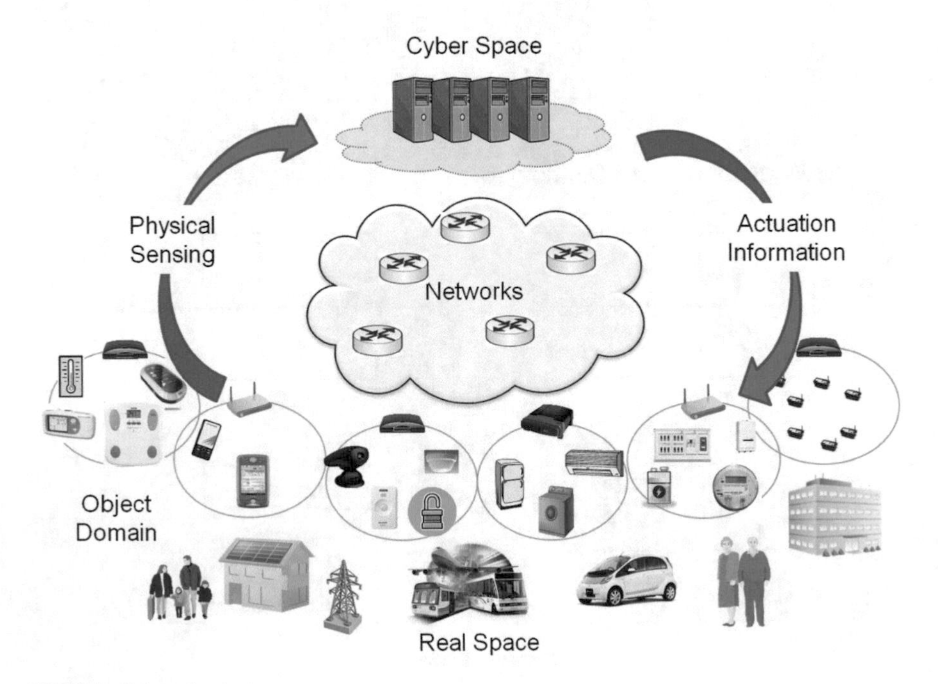

Fig. 8.3 Cyber physical systems

lying communication infrastructure. For example, CPS applications (see Fig. 8.3) are a design paradigm for engineering real-life applications [52–57] that controlled and monitored by computer-based algorithms, tightly integrated with Internet and its users. In CPS, physical and software components are deeply intertwined, each operating on different spatial and temporal scales, exhibiting multiple and distinct behavioral modalities, and interacting with each other in a myriad of ways that change with context. Real-life examples of CPS are smart grids, SCADA systems, etc.

Similarly, the STS is an approach to complex organizational work design that recognizes the interaction between people and technology in workplaces. The term also refers to the interaction between society's complex infrastructures and human behavior. Both of CPS and STS are instances of CAS (Complex Adaptive Systems) which are fluidly changing collections of distributed interacting components that react to both their environments and to one another. The engineering of cyberspace applications such as CPS and STS is currently a challenging task that has a great focus worldwide because having adaptive, robust, secure, and scalable cyberspace applications will have a critical role not only during peace times but also during wars as they enable decision makers to rapidly, effectively, and rationally taking critical decisions. Traditionally, the development of cyberspace applications relied on conventional web technologies (see Fig. 8.4a) such as web browsers, web servers, web services, applets, servlets, etc. These conventional approaches used the client-server architecture which adopts the centralized control approach and widely supported as a method provide a simple mean to develop distributed applications. However, it is more suited to centralized applications, in which one server serves a number of clients, or one client controls a number of servers, than true complex highly distributed applications, and is lacking in flexibility and robustness and cant easily accommodate new requirements or demands of modern large-scale complex real-life applications. Recently, the agents and Multiagent system new software engineering paradigm emerged as a feasible and effective approach that (MAS) adopts the decentralized control paradigm. It is being promoted by the software engineering community to be the adequate solution to handle the current requirements of complex engineering problems that demanding distribution, flexibility, robustness and dynamic re-reconfigurability. Intuitively, autonomy, adaptivity, robustness, and scalability, are generally considered as the key properties that should be realized within modern complex, large-scale and highly distributed systems and these are exactly the properties that characterize MAS [58].

Using MAS for developing modern cyberspace applications can effectively enable building adaptive applications capable of efficiently adapting to failures, component replacements and changes in the environment with less human intervention or centralized management. Intelligent information agents can be deployed globally within the Internet and interact together to achieve intelligently and effectively the designed application objectives (Fig. 8.4b). The agents can be prepared to interact with non-agent objects using proprietary or standard interfaces. On the other hand, for the sake of scalability and security it is better to agentifing (make them as agents) the non-agent objects so that all the communication themes are done based on standard Agent Communication Language (ACL) [59] and proper domain ontologies [60],

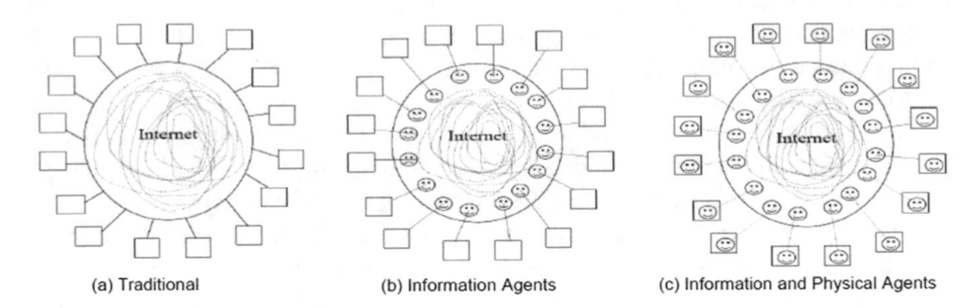

Fig. 8.4 The evolution of internet-based applications

Fig. 8.5 Agent-based modeling of modern real-life applications

(Fig. 8.4c). From the other hand, the problem with the adoption of the agent-based approach for engineering modern real-life cyberspace applications will appear when there exist a very large number of agents in one application. How can agents coordinate their activities? how can they find each other within the Internet ocean? Further, how to manage the complexity of agents interactions? To demonstrate these problems and challenges, consider Fig. 8.5 which presents the consequences of modeling and engineering CPS and STS using the MAS paradigm. Originally, this type of systems is characterized by highly distribution, openness, heterogeneity, complexity, unpredictable and uncertain environment, and their Large-scale nature, and after modeling them using the agent-based approach, the result will be a large number of agents interacting with their environments and with each other through cyberspace (Internet-Agents).

The problems that can appear are:

- Complexity: complex interactions and emergent behaviors.
- Coordination: how to align the behavior of different agents.
- Agent location: how can agents find each other to be able to interact?

- Uncertainty: imperfect and unknown information can be found within these systems.
- Big Data: These systems may include data sets with sizes beyond the ability of commonly used software tools to capture.

These problems can be handled by providing a way to statically and dynamically structuring and restructuring the MAS, respectively. By structuring and restructuring we mean statically organizing and dynamically reorganizing the MAS (in run-time as an adaptive behavior) as a response to environment dynamic changes. In the proposed framework, we adopt a recent model for dynamically reorganizing the application as an adaptive behavior.

Another important challenge that is facing modern cyberspace applications is cyberspace security. The increasing usage of Internet as a communication medium for this type of emerging applications made them more vulnerable to a huge number of threats and cyber-attacks. Making cyberspace secure and trustworthy is one of the most important challenges confronting modern cyber applications. As MAS allow the design and implementation of software systems using the same ideas and concepts that are the very founding of human societies and habits. The agent-based approach enables the use of soft security methods such as the trust, reputation, gossip, etc. methods in addition to the hard security methods such as authentication, encryption, firewalls, etc. The proposed project aims to provide a novel trust-based model for securing cyberspace agent-based applications such as CPS. Another challenge is the forensic investigation of modern real-life cyberspace applications in case of catastrophic security incidents. Forensic investigators have the responsibility to determine the main causes of the catastrophic incidents that might happen and provide precise and logical evidences supported with comprehensive technical reports to the legal organizations. Unfortunately, they are facing many obstacles against the accomplishment of their critical mission. These obstacles are mainly related to the availability of the technical process data in case of catastrophic incidents. Examples of these obstacles are: live process data acquisition, highly distributed locations, multiple operational levels, and many other challenges [61]. The proposed project aims to enable the forensic investigators to do their mission from anywhere, and at any time using local (in-site) or enterprise communication networks such as the Internet. Also, differing from other related approaches which mainly relied on post-incident data, the proposed project aims to use the pre-, in-, and post-incident data to precisely discover the causes of an incident. WSN are spatially distributed autonomous sensors to monitor physical or environmental conditions, such as temperature, sound, pressure, etc. They are powerful in providing real-time (live) information about the work environments of CPS and STS. The proposed project will also use WSN to help forensically analyzing and discovering the real causes of cyberspace incidents. In summary, this research project aims to provide a holistic framework for securely and forensically engineering adaptive cyberspace real-life applications such as CPS and STS based on MAS and WSN.

8.4 The Proposed Model

To be self-content, we briefly define SCADA systems components. A SCADA System usually consists of a set of subsystems. A Human-Machine Interface (HMI) which presents process data to a human operator to monitors and controls the process. A supervisory (computer) system, gathering (acquiring) data on the process and sending commands (control) to the process. This system uses a Remote Terminal Units (RTUs) connecting to sensors in the process to convert sensor signals to digital data and sending digital data to the supervisory system. Programmable Logic Controller (PLCs) used as field devices because they are more economical, versatile, flexible, and configurable than special purpose RTUs. Communication infrastructure connects the supervisory system to the RTUs.

The proposed method aims to use WSN to acquire live data on a SCADA system. Using WSN enables us to acquire this data in a way that minimizes risk to the systems services. This can be achieved by distribute a set of sensor nodes through the working field of the SCADA system. The count of the distributed sensor will depend on the volatile data at each component of the SCADA system. Sensors are able to transmit data as a response to an event, i.e. change in volatile data, or a request from BS. The sensors will send the collected data to the base station (BS) which is connected with a processing server that is able to manage these data using a file checker software for file processing (read, write, save, delete, modify, and generate reports).

Thus, we minimize the risk to the systems services by using WSN to frequently send volatile data abroad out of the system to the BS to process it on a separate server. The time between rounds of data transmission depends on the changes on the volatile data. If the file checker tool stored at the server finds any change in the file integrity roles that are predefined by the system, the transferred copy of the predefined data will be stored. Then, the BS will keep sending request to sensors to collect volatile data from all devices in order to save all changes before, during and after incident is happened. At the BS side, If the server finds anything wrong in the coming data at time (t), the BS will keep sending request to all sensors to collect all available data as possible. Then, the collected data at time t, $t + 1$, $t + 2$, $t + 3, \ldots$ will be stored at the server to be used by the Forensic investigators.

Following our proposed method, forensic investigators can avoid a lot of challenges that face them after an incident occurred. For example, the proposed method guarantees early data acquisition after incident. That is because; the volatile data is captured as quickly as possible if any changes occurred. The BS will save the data as soon as it received it. Moreover, the proposed system will provide digital evidence validity by storing a copy of the volatile data at BS before and after an incident. Thus, the Forensic investigators can use these data to prove the integrity of evidence.

Finally, some of the strengths of the proposed method can be listed as the following:

- There is no effect on the system services while Forensic investigators doing their work.

- WSN is used for both monitoring and acquiring data from SCADA.
- WSN is separate and in not affected by any attack of SCADA system.
- WSN uses cryptography to transfer data from sensors to the BS (at which the data is processed).
- Forensic investigators can use the data stored at BS to do the required analysis. The connected server at the BS can runs an intelligent tool to collect a specific data at specific time to provide an automated report to the Forensic investigators.
- WSN is inexpensive to setup and all type of sensors are available. The sensors will be distributed based on the number of devices that contains volatile data. In other word, sensors will be distributed based on the architecture of the SCADA system.
- As we know, the main challenge of WSN is the energy source. However, in our system, each sensor has a static and predefined location. So, we can provide every sensor by power supply to recharge it periodically.
- WSN provides effective way to cover a wide area field.
- The proposed system can work with any existing SCADA system regardless its architecture.
- The proposed system represents a second line of data security in the SCADA environment. However, there is no need to change the security procedures applied at SCADA.

8.5 The Proposed Models Architecture and Framework

As shown in Fig. 8.6, the proposed architecture of the SCADA system works by distributing a set of sensors around the working SCADA system. These sensors transmit data outside of the SCADAs working field to the data collecting point (BS). The BS is connected with a data processing server which manages all incoming data and queries. The data is stored in a data repository to allow to a forensic investigator to get it when he/she needs. The advantage of this architecture is its simplicity. That is because there is no need to modify the existing SCADA architecture. In addition, there are a lot of common tools and software that can be used to manage the data at the data processing servers side.

In case of distributed SCADA system that covers wide area, the cluster model of WSN can be used. Each cluster will be responsible of collecting the data from a distributed part of the SCADA system and transfer if directly to the BS as shown at Fig. 8.7.

We can use The MICAz sensors which are based on the low-power 8-bit microcontroller ATmega128L with a clock frequency of 7.37 MHz and runs TinyOS as an event driven operating system. It also embeds a IEEE 802.15.4 compliant CC2420 transceiver with a claimed data rate of 250 kbps.

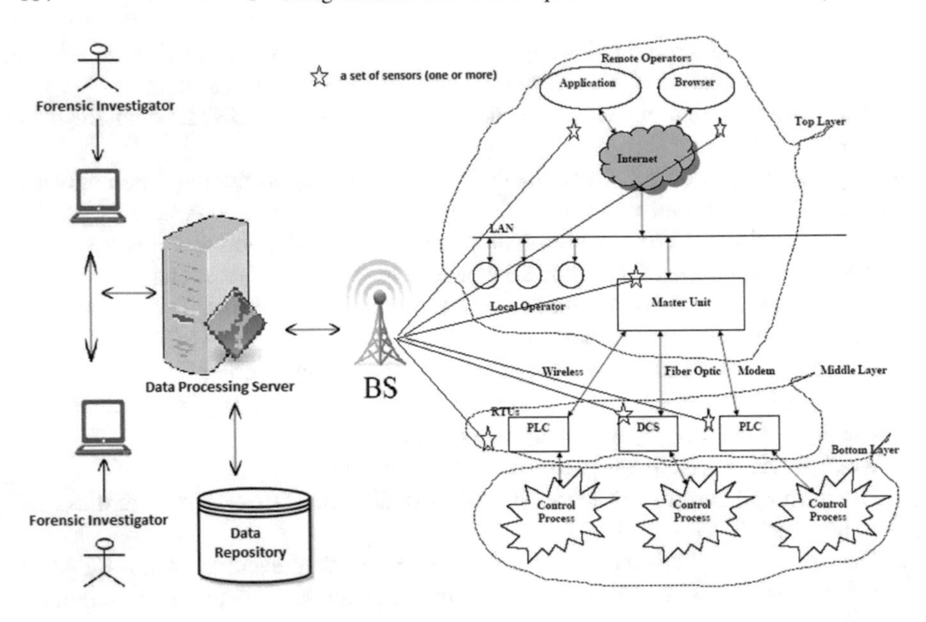

Fig. 8.6 The proposed architecture of the SCADA system

Fig. 8.7 The WSN clustering model to be used with distributed SCADA system

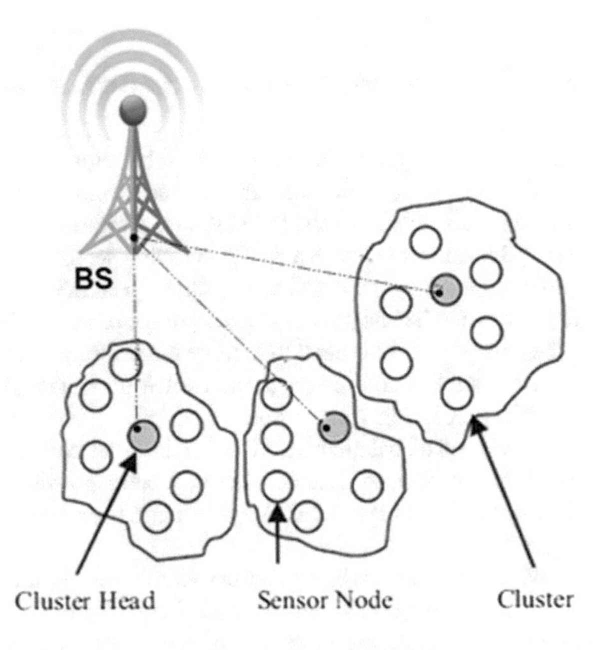

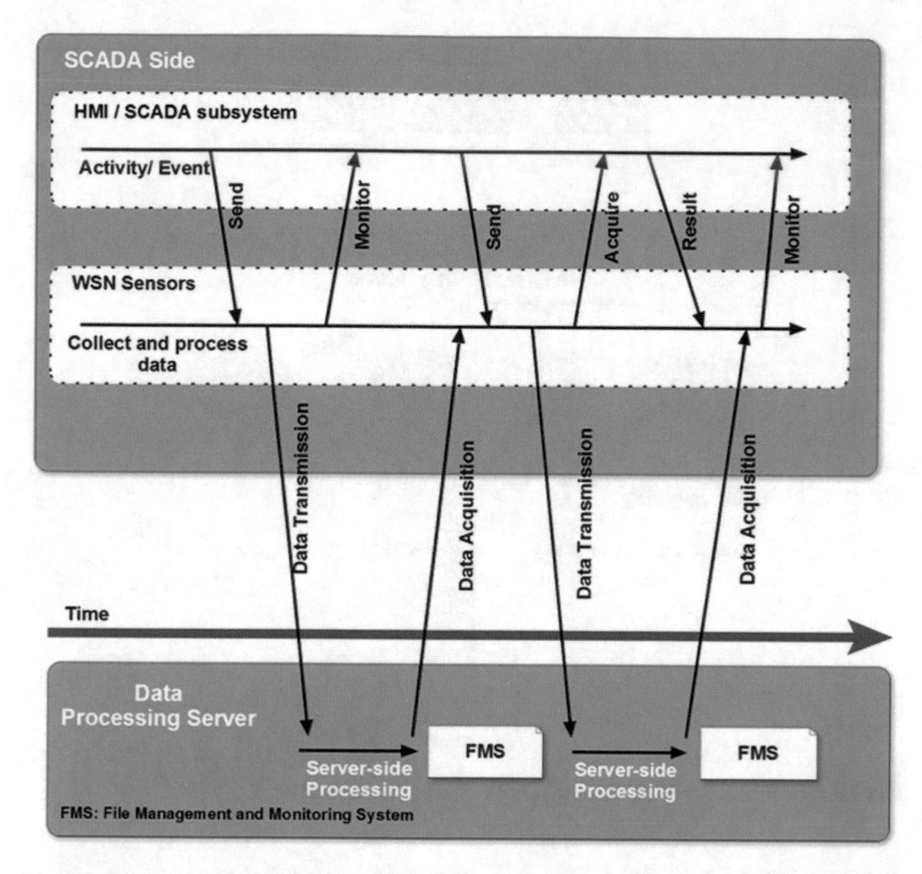

Fig. 8.8 The synchronizing process between a SCADA and the data processing server

As shown in Fig. 8.8, the data transmission process is based on a synchronous model between a SCADA system side and the data processing servers side. The SCADA system side includes two main components: (1) SCADA subsystem at which the event is occurred, and (2) the WSN sensors which are distributed to monitor all events and collect data. On the other side, the data processing server side responsible for data processing and file integrity monitoring using the file management and monitoring system (FMS).

Figure 8.9 shows the FMSs interaction with the data processing server. This interaction initiates by request from forensic investigator who needs to get data from the data repository. FMS works as the intermediate between the stored data and the investigators.

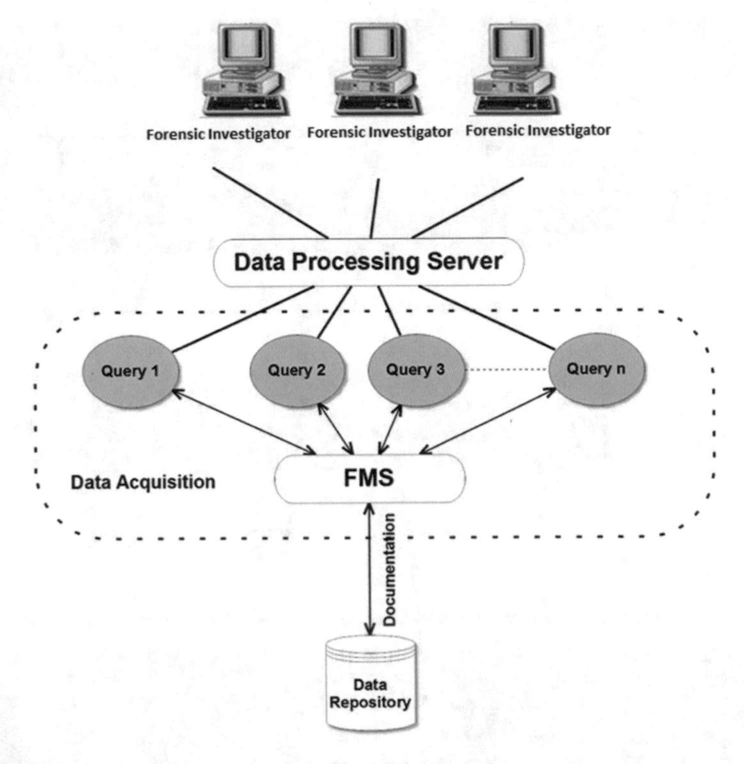

Fig. 8.9 The interaction between FMS and the data processing server

8.6 The Working Steps Towards Implement the Proposed Method

In order to implement the proposed method, a set of steps must be followed. These steps depend on a set of parameters such as the working field, the installed SCADA system and its working procedures, the type of data, integration tools between WSN and SCADA components, and existing software tools as well. So, it is required to setup a sequence of phases to follow when trying to build our method as shown in Fig. 8.10.

8.7 Conclusion

SCADA system is a computer system spread over a wide area in order to remotely control and monitoring a specific process in a working field. Since SCADA systems are network presence systems and face significant threats and vulnerabilities, many algorithms that try to maximize the security procedures of a SCADA system have

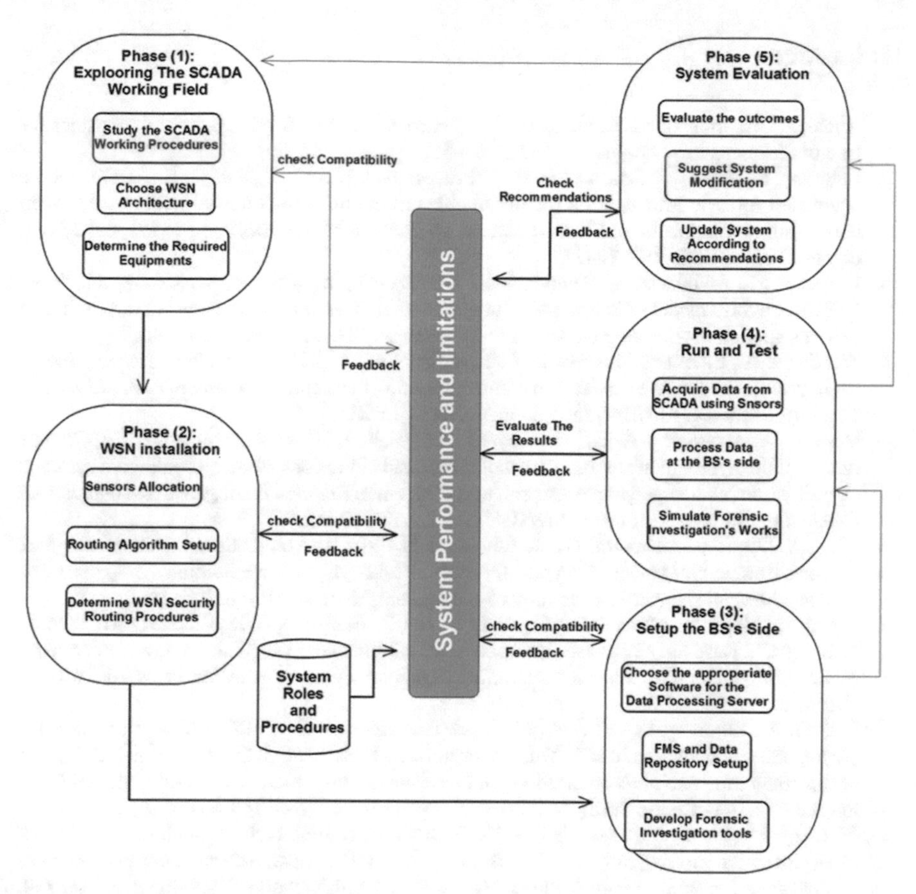

Fig. 8.10 The working steps towards implement the proposed method

been developed. But little is available that deals specifically with live data acquisition for digital forensic. However, volatile data acquisition still represents a big challenge for the researchers because it may affect the system performance. Therefore, we propose a new method for live data acquisition using WSN to acquire live data without affecting the SCADA systems performance. The proposed idea attempts to monitor file integrity and acquire live data in a way that minimizes risk to the systems services. In addition, it aims to help Forensic investigators by guarantee early data acquisition after incident and digital evidence validity as well. The proposed method depends on distributing a set of sensor nodes across the SCADAs working field to capture the volatile data. The objectives, strengths, architecture, and implementation details of the proposed method were discussed in this proposal.

References

1. Ahmed, I., Obermeier, S., Naedele, M., & Richard, G. (2012). Scada systems: Challenges for forensic investigators. *Computer*, *45*(12), 44–51.
2. Elhoseny, M., Hosny, A., Hassanien, A. E., Muhammad, K., & Sangaiah, A. K. (2017). Secure automated forensic investigation for sustainable critical infrastructures compliant with green computing requirements. *IEEE Transactions on Sustainable Computing*, *PP*(99). https://doi.org/10.1109/TSUSC.2017.2782737.
3. Elhoseny, M., Abdelaziz, A., Salama, A. S., Riad, A. M., Muhammad, K., & Sangaiah, A. K. (2018). A hybrid model of internet of things and cloud computing to manage big data in health services applications. *Future Generation Computer Systems*. Elsevier. (in Press).
4. Abdelaziz, A., Elhoseny, M., Salama, A. S., & Riad, A. M. (2018). A machine learning model for improving healthcare services on cloud computing environment. *Measurement*, *119*, 117–128. https://doi.org/10.1016/j.measurement.2018.01.022.
5. Darwish, A., Hassanien, A. E., Elhoseny, M., Sangaiah, A. K., & Muhammad, K. (2017). The impact of the hybrid platform of internet of things and cloud computing on healthcare systems: Opportunities, challenges, and open problems. *Journal of Ambient Intelligence and Humanized Computing*. Springer. https://doi.org/10.1007/s12652-017-0659-1.
6. Yuan, X., Li, D., Mohapatra, D., & Elhoseny, M. (2017). Automatic removal of complex shadows from indoor videos using transfer learning and dynamic thresholding. *Computers and Electrical Engineering*. https://doi.org/10.1016/j.compeleceng.2017.12.026. (in Press).
7. Sajjad, M., Nasir, M., Muhammad, K., Khan, S., Jan, Z., Sangaiah, A. K., Elhoseny, M., & Baik, S. W. (2017). Raspberry Pi assisted face recognition framework for enhanced law-enforcement services in smart cities. *Future Generation Computer Systems*. Elsevier. https://doi.org/10.1016/j.future.2017.11.013.
8. Shehab, A., Elhoseny, M., El Aziz, M. A., & Hassanien A. E. (2018). Efficient schemes for playout latency reduction in P2P-VoD systems. In A. Hassanien, & D. Oliva (Eds.), *Advances in soft computing and machine learning in image processing*. Studies in Computational Intelligence, Vol. 730. Cham: Springer. https://doi.org/10.1007/978-3-319-63754-9_22.
9. Elhoseny, M., Nabil, A., Hassanien, A. E., & Oliva, D. (2018). Hybrid rough neural network model for signature recognition. In A. Hassanien, & D. Oliva (Eds.) *Advances in soft computing and machine learning in image processing*. Studies in Computational Intelligence, Vol. 730. Cham: Springer. https://doi.org/10.1007/978-3-319-63754-9_14.
10. Abdeldaim A. M., Sahlol A. T., Elhoseny M., & Hassanien A. E. (2018). Computer-aided acute lymphoblastic Leukemia diagnosis system based on image analysis. In: A. Hassanien, D. Oliva (Eds.), *Advances in soft computing and machine learning in image processing*. Studies in Computational Intelligence, Vol. 730. Cham: Springer. https://doi.org/10.1007/978-3-319-63754-9.
11. Abbas, H. (2014). Future SCADA challenges and the promising solution: The agent-based SCADA. *International Journal of Critical Infrastructures*, *10*(3/4), 307–333.
12. Elhoseny, M., Ramírez-González, G., Abu-Elnasr, O. M., Shawkat, S. A., Arunkumar, N., & Farouk, A. (2018). Secure medical data transmission model for IoT-based healthcare systems. *IEEE Access*, *PP*(99). https://doi.org/10.1109/ACCESS.2018.2817615.
13. Shehab, A., Elhoseny, M., Muhammad, K., Sangaiah, A. K., Yang, P., Huang, H., & Hou, G. (2018). Secure and robust fragile watermarking scheme for medical images. *IEEE Access*, *6*(1), 10269–10278. https://doi.org/10.1109/ACCESS.2018.2799240.
14. Farouk, A., Batle, J., Elhoseny, M., Naseri, M., Lone, M., Fedorov, A., Alkhambashi, M., Ahmed, S. H., & Abdel-Aty, M. (2018). Robust general N user authentication scheme in a centralized quantum communication network via generalized GHZ states, *Frontiers of Physics*, *13*, 130306. Springer. https://doi.org/10.1007/s11467-017-0717-3.
15. Elhoseny, M., Elkhateb, A., Sahlol, A., & Hassanien, A. E. (2018). Multimodal biometric personal identification and verification. In A. Hassanien, & D. Oliva (Eds.), *Advances in soft computing and machine learning in image processing*. Studies in Computational Intelligence, Vol. 730. Cham: Springer. https://doi.org/10.1007/978-3-319-63754-9_12.

16. Elhoseny, M., Essa, E., Elkhateb, A., Hassanien, A. E., & Hamad, A. (2018). Cascade multimodal biometric system using fingerprint and Iris patterns. In A. Hassanien, K. Shaalan, T. Gaber, & M. Tolba (Eds.), *Proceedings of the international conference on advanced intelligent systems and informatics 2017*, AISI 2017. Advances in Intelligent Systems and Computing, Vol. 639. Cham: Springer. https://doi.org/10.1007/978-3-319-64861-3_55.
17. Wu, T., Disso, J. F. P., Jones, K., & Campos, A. (2013). Towards a SCADA forensics architecture. In *Proceedings of the 1st international symposium for ICS and SCADA cyber security research*, (pp. 12–21).
18. Spyridopoulos, T., Tryfonas, T., & May, J. (2014). Incident analysis & digital forensics in SCADA and industrial control systems. In *8th IET international system safety conference incorporating the cyber security*. IEEE.
19. Pedro, N. (2013). SCADA live forensics: real time data acquisition process to detect, prevent, or evaluate critical situations. In *1st annual international interdisciplinary conference*, (pp. 24–26).
20. Tharwat, A., Mahdi, H., Elhoseny, M., & Hassanien, A. E. (2018). Recognizing human activity in mobile crowdsensing environment using optimized k-NN algorithm. *Expert Systems With Applications*. https://doi.org/10.1016/j.eswa.2018.04.017. Accessed 12 April 2018.
21. Tharwat, A., Elhoseny, M., Hassanien, A. E., Gabel, T., & Kumar, A. (2018). Intelligent Bezir curve-based path planning model using chaotic particle swarm optimization algorithm. *Cluster Computing*, (pp. 1–22). Springer. https://doi.org/10.1007/s10586-018-2360-3.
22. Sarvaghad-Moghaddam, M., Orouji, A. A., Ramezani, Z., Elhoseny, M., & Farouk, A. (2018). Modelling the spice parameters of SOI MOSFET using a combinational algorithm. *Cluster Computing*. Springer. https://doi.org/10.1007/s10586-018-2289-6. (in Press).
23. Rizk-Allah, R. M., Hassanien, A. E., & Elhoseny, M. (2018). A multi-objective transportation model under neutrosophic environment. *Computers and Electrical Engineering*. Elsevier. https://doi.org/10.1016/j.compeleceng.2018.02.024. (in Press).
24. Batle, J., Naseri, M., Ghoranneviss, M., Farouk, A., Alkhambashi, M., & Elhoseny, M. (2017). Shareability of correlations in multiqubit states: Optimization of nonlocal monogamy inequalities. *Physical Review A*, *95*(3), 032123. https://doi.org/10.1103/PhysRevA.95.032123.
25. El Aziz, M. A., Hemdan, A. M., Ewees, A. A., Elhoseny, M., Shehab, A., Hassanien, A. E., & Xiong, S. (2017). Prediction of biochar yield using adaptive neuro-fuzzy inference system with particle swarm optimization. In *IEEE PES PowerAfrica conference*, (pp. 115–120), June 27–30, 2017. Accra-Ghana: IEEE. https://doi.org/10.1109/PowerAfrica.2017.7991209.
26. Ewees, A. A., El Aziz, M. A., & Elhoseny, M. (2017) Social-spider optimization algorithm for improving ANFIS to predict biochar yield. In *8th International conference on computing, communication and networking technologies (8ICCCNT)*, July 3–5. Delhi-India: IEEE.
27. Metawa, N., Elhoseny, M., Hassan, M. K., & Hassanien, A. E. (2016). Loan portfolio optimization using genetic algorithm: A case of credit constraints. In *Proceedings of 12th international computer engineering conference (ICENCO)*, (pp. 59–64). IEEE. https://doi.org/10.1109/ICENCO.2016.7856446.
28. Elhoseny, M., Farouk, A., Batle, J., Shehab, A., & Hassanien, A. E. (2017). Secure image processing and transmission schema in cluster-based wireless sensor network. In *Handbook of research on machine learning innovations and trends*, Chapter 45, pp. 1022–1040, IGI Global, 2017. https://doi.org/10.4018/978-1-5225-2229-4.ch045.
29. Elhoseny, M., Elleithy, K., Elminir, H., Yuan, X., & Riad, A. (2015). Dynamic clustering of heterogeneous wireless sensor networks using a genetic algorithm towards balancing energy exhaustion. *International Journal of Scientific & Engineering Research*, *6*(8), 1243–1252.
30. Yuan, X., Elhoseny, M., El-Minir, H., & Riad, A. (2017). A genetic algorithm-based, dynamic clustering method towards improved wsn longevity. *Journal of Network and Systems Management*, *25*(1), 21–46.
31. Elhoseny, M., Yuan, X., Yu, Z., Mao, C., El-Minir, H., & Riad, A. (2015). Balancing energy consumption in heterogeneous wireless sensor networks using genetic algorithm. *IEEE Communications Letters*, *19*(12), 2194–2197. IEEE. https://doi.org/10.1109/LCOMM.2014.2381226.

32. Elhoseny, M., Tharwat, A., Yuan, X., & Hassanien, A. E. (2018). Optimizing K-coverage of mobile WSNs. *Expert Systems with Applications*, *92*, 142–153. Elsevier. https://doi.org/10.1016/j.eswa.2017.09.008.
33. Elhoseny, M., Tharwat, A., Farouk, A., & Hassanien, A. E. (2017). K-coverage model based on genetic algorithm to extend WSN lifetime. *IEEE Sensors Letters*, *1*(4), 1–4. IEEE. https://doi.org/10.1109/LSENS.2017.2724846.
34. Elhoseny, M., Farouk, A., Zhou, N., Wang, M. M., Abdalla, S., & Batle, J. (2017). Dynamic multi-hop clustering in a wireless sensor network: Performance improvement. *Wireless Personal Communications*, *95*(4), 3733–3753. Springer. https://doi.org/10.1007/s11277-017-4023-8.
35. Elhoseny, M., Yuan, X., El-Minir, H. K., & Riad, A. (2014). Extending self-organizing network availability using genetic algorithm. In *International Conference on Computing, Communication and Networking Technologies (ICCCNT)*, (pp. 1–6). IEEE.
36. Yuan, X., Elhoseny, M., El-Minir, H. K., & Riad, A. M. (2017). A genetic algorithm-based, dynamic clustering method towards improved WSN longevity. *Journal of Network and Systems Management*, *25*(1), 21–46. Springer. https://doi.org/10.1007/s10922-016-9379-7.
37. Elhoseny, M., Yuan, X., El-Minir, H. K., & Riad, A. M. (2016b). An energy efficient encryption method for secure dynamic WSN. *Security and Communication Networks*, *9*(13), 2024–2031.
38. Elhoseny, M., Elminir, H., Riad, A., & Yuan, X. (2016a). A secure data routing schema for WSN using elliptic curve cryptography and homomorphic encryption. *Journal of King Saud University-Computer and Information Sciences*, *28*(3), 262–275.
39. Elsayed, W., Elhoseny, M., Riad, A., & Hassanien, A. E. (2017). Autonomic self-healing approach to eliminate hardware faults in wireless sensor networks. In *International conference on advanced intelligent systems and informatics*, (pp. 151–160). Springer.
40. Elsayed, W., Elhoseny, M., Sabbeh, S., & Riad, A. (2017). Self-maintenance model for wireless sensor networks. *Computers and Electrical Engineering*. https://doi.org/10.1016/j.compeleceng.2017.12.022. (in Press).
41. Elhoseny, M., Yuan, X., El-Minir, H. K., & Riad, A. M. (2016). An energy efficient encryption method for secure dynamic WSN. *Security and Communication Networks*, *9*(13) 2024–2031. https://doi.org/10.1002/sec.1459.
42. Hosseinabadi, A. A. R., Vahidi, J., Saemi, B., Sangaiah, A. K., & Elhoseny, M. (2018). Extended genetic algorithm for solving open-shop scheduling problem. *Soft Computing*. https://doi.org/10.1007/s00500-018-3177-y.
43. Abbas, H. A. (2014). Exploiting the overlapping of higher order: Entities within multi-agent systems. *International Journal of Agent Technologies and Systems (IJATS)*, *6*(3), 32–57.
44. Abbas, H. A. (2015). Realizing the NOSHAPE MAS Organizational model: An operational view. *International Journal of Agent Technologies and Systems (IJATS)*, *7*(2), 75–104.
45. Abbas, H. A., Shaheen, S. I., & Amin, M. H. (2016). Self-adaptive large-scale SCADA system based on self-organised multi-agent systems. *International Journal of Automation and Control*, *10*(3), 234266.
46. Bellifemine, F., Poggi, A., & Rimassa, G. (1999). JADE: A FIPA-compliant agent framework. In *Proceedings of the practical applications of intelligent agents and multi-agents*, (pp. 97–108).
47. Foundation for Intelligent Physical Agents (FIPA) (2000) FIPA Agent Management Specification. http://www.fipa.org/specs/fipa00023/.
48. Moreno, A., Valls, A., & Viejo, A. (2003). Using JADE-LEAP to Implement Agents in Mobile Devices. http://jade.tilab.com/papers/EXP/02Moreno.pdf.
49. Braubach, L., Pokahr, A., Bade, D., Krempels, K. H., & Lamersdorf, W. (2004). Deployment of distributed multi-agent systems. In *International workshop on engineering societies in the agents world*, (pp. 261–276). Heidelberg: Springer.
50. Saqib, A., Anwar, R. W., Hussain, O. K., Ahmad, M., Ngadi, M. A., Mohamad, M. M., Malki, Z. O. H. A. I. R., Noraini, C., Jnr, B. A., Nor, R. N. H. & Murad, M. A. A. (2015). Cyber security for cyber physcial systems: a trust-based approach. *Journal of theoretical and applied information technology*, *71*(2).

51. Neuman, C., & Tan, K. (2011). Mediating cyber and physical threat propagation in secure smart grid architectures. *IEEE International Conference on Smart Grid Communications, 17–20*, 238243.
52. Elhoseny, H., Elhoseny, M., Riad, A. M., Hassanien, A. E. (2018). A framework for big data analysis in smart cities. In A. Hassanien, M. Tolba, M. Elhoseny, M. Mostafa (Eds.), *AMLTA 2018 the international conference on advanced machine learning technologies and applications (AMLTA2018)*. Advances in Intelligent Systems and Computing, Vol. 723. Cham: Springer. https://doi.org/10.1007/978-3-319-74690-6_40.
53. Elhoseny M., Shehab A., & Osman L. (2018) An empirical analysis of user behavior for P2P IPTV workloads. In A. Hassanien, M. Tolba, M. Elhoseny, & M. Mostafa (Eds.) *AMLTA 2018 The International Conference on Advanced Machine Learning Technologies and Applications (AMLTA2018)*. Advances in Intelligent Systems and Computing, Vol. 723. Cham: Springer https://doi.org/10.1007/978-3-319-74690-6_25.
54. Wang, M. M., Qu, Z. G., Elhoseny, M. (2017). Quantum secret sharing in noisy environment. In X. Sun, H. C. Chao, X. You, & E. Bertino (Eds.) *Cloud computing and security, ICCCS 2017*. Lecture Notes in Computer Science, Vol. 10603. Cham: Springer. https://doi.org/10.1007/978-3-319-68542-7_9.
55. Elsayed, W., Elhoseny, M., Riad, A. M., & Hassanien, A. E. (2018). Autonomic self-healing approach to eliminate hardware faults in wireless sensor networks. In A. Hassanien, K. Shaalan, T. Gaber, & M. Tolba (Eds.), *Proceedings of the international conference on advanced intelligent systems and informatics 2017*, AISI 2017. Advances in Intelligent Systems and Computing, Vol. 639. Cham: Springer. https://doi.org/10.1007/978-3-319-64861-3_14.
56. Abdelaziz, A., Elhoseny, M., Salama, A. S., Riad, A. M., & Hassanien, A. E. (2018). Intelligent algorithms for optimal selection of virtual machine in cloud environment, towards enhance healthcare services. In A. Hassanien, K. Shaalan, T. Gaber, & M. Tolba (Eds.), *Proceedings of the international conference on advanced intelligent systems and informatics 2017*, AISI 2017. Advances in Intelligent Systems and Computing, Vol. 639. Cham: Springer. https://doi.org/10.1007/978-3-319-64861-3_27.
57. Shehab, A., Ismail, A., Osman, L., Elhoseny, M., El-Henawy, I. M. (2018). Quantified self using IoT wearable devices. In A. Hassanien, K. Shaalan, T. Gaber, M. Tolba (Eds.), *Proceedings of the international conference on advanced intelligent systems and informatics 2017*, AISI 2017. Advances in Intelligent Systems and Computing, Vol. 639. Cham: Springer. https://doi.org/10.1007/978-3-319-64861-3_77.
58. Weyns, D., Helleboogh, A., & Holvoet, T. (2009). How to get multi-agent systems accepted in industry? *International Journal of Agent-Oriented Software Engineering*, *3*(4), 383–390.
59. Foundation For Intelligent Physical Agents (1997), Agent Communication Language, FIPA 97 Specification Part 2.
60. Annamalai, M., & Sterling, L. (2003). Guidelines for constructing reusable domain ontologies. In *OAS*, (pp. 71–74).
61. Ahmed, I., Obermeier, S., Naedele, M., & Richard III, G. G. (2012). SCADA systems: Challenges for forensic investigators. *Computer*, *45*(12), 44–51.

Printed in the United States
By Bookmasters